5 Steps Chemistry Workbook Series
Book 1: Sassy Stoichiometry Problems

14 Worksheets and Keys for College Prep and Honors Chemistry High School Students

By Julie C. Gilbert

Aletheia Pyralis Publishers

For information about special discounts available for bulk purchases, sales promotions, fund-raising and educational needs, please email: juliecgilbert5steps@gmail.com.

http://www.juliecgilbert.com/

Table of Contents:

Introduction:

What is this book and how do I use it?

This will be a collection of worksheets designed for college prep and honors chemistry students to practice stoichiometry. Each section will have the "easier" practice sets first.

For convenience, I will likely be providing a list of the atomic masses that will be used in this book because every periodic table will round to a different number of decimal places. For testing and quizzing purposes, you will likely have to use the periodic table provided by your teacher.

Tip: If your teacher allows it, you may wish to write out a list of common atomic masses you will need. With enough practice, you'll likely memorize the important ones, but there are several elements that show up in a lot of compounds: hydrogen, oxygen, carbon, nitrogen, sulfur, phosphorus.

Tip: My keys and explanations will be designed around one of the most basic periodic tables I can find because I'd rather not drown in decimals along the way. However, you should solve the problems with your class periodic table. I also recommend using the calculator you intend to use during assessments. The more you use the normal tools at your disposal, the more comfortable you will be with the process.

Why should I use this book when I can find a million articles and a dozen videos online?

When it comes to mastering chemistry concepts, there's really no substitute for practice. Please do find some videos that cover the topic and use them to guide you. This will be a book of worksheets with the keys. I absolutely love audiobooks and ebooks are okay too, but for this, I highly recommend the paperback because the main point is for each student to be able to physically write out the answers.

Notes about the Worksheets in this Book:

Note 1: I am not providing molar masses on the first worksheet because you will not need them. However, I will go through the mole ratios because that is the focus of the first worksheet. Many worksheets have the molar masses, but you should still practice calculating them on your own.

Note 2: Do not reduce mole ratios. They come directly from the balanced equation. If you get in the habit of reducing, it'll be easier to lose track of where these numbers came from.

Note 3: All flipped versions of mole ratios are also acceptable, but I don't list them because they're duplicates and not really new relationships, merely a different way of looking at the same relationship.

Note 4: The typed fractions do not format the subscripts properly. You should use proper subscripts when writing chemical formulas.

Note 5: The wording of the question gives you clues as to whether the chemicals talked about are reactants or products. Phrases like "are needed" and "react to form" indicate that the chemical is a reactant. Phrases like "are formed" and "are produced" indicate that the chemical being discussed is a product.

Note 6: I will sometimes change the wording of the questions and explanations because there's no way to predict the way your teacher will write questions, and sometimes, a particular wording will work for a student where another one fails to connect. It's best if you see a variety of ways to word problems and explanations. Same reasoning for why some explanations are written as bullet points and others are full paragraphs.

Note 7: For the same reason as note 6, I will occasionally switch up the number of significant figures. Always read the directions in a particular worksheet. I might even decide you should just round to the hundredths for a certain worksheet. I've done that with classes where I'm not concerned about significant figures.

Note 8: The answer keys will be immediately after each worksheet.

Note 9: General Directions: Balance the equation and calculate molar masses where needed. Use the conversion factor method. Keep all units and labels throughout each problem. Watch significant figures unless otherwise instructed.

Background and Skills for Mastering Stoichiometry:

What is Stoichiometry?

Stoichiometry relates the parts of a balanced chemical equation to one another. If you start with a certain amount of reactants, stoichiometry will allow you to calculate how much of the product(s) you should be able to get from the reaction.

Disclaimer: If you've read *5 Steps to Surviving Chemistry*, some of this will be familiar to you. I'm going to recycle the relevant parts so people can review before trying the practice sets.

Skills you need to do Stoichiometry:

- Balance chemical equations
- Find the average atomic mass on your assigned periodic table
- Calculate molar mass
- Use the balanced equations to find relevant mole ratios
- Round answers to proper significant figures

Helpful hints to balancing equations:

1. Balance the different types of atoms one by one
2. First balance atoms that only show up once on each side
3. Balance polyatomic ions that appear on both sides as a single unit (you may use a substitution method if it is easier for you)
4. Balance H and O after all other atoms. These typically show up the most, thus are most difficult to balance. As a general rule of thumb, if an element is in more than one place on a given side of the equation, save it for last.

Key point: "balanced" means there are the same number of each element on each side

Steps to calculating formula mass and molar mass:
Step #1: Write the balanced chemical formula for the compound
Step #2: Write down the number of each type of atom
Step #3: Multiply the number of each type of atom by its average atomic mass. (These values are found on the periodic table of elements.)
Step #4: Add the values together and report the final answer.

Note: Molar mass is obtained the same way. The only difference is that the units along the way will be grams and the final unit will be g/mol.

Dimensional Analysis (Conversion Factor Method):

Conversion Factor: a way of presenting a particular equivalent value. For example, 1 dollar = 4 quarters. Think of it as an equivalency expressed as a fraction. That might sound very fancy, but it simply means that the numerator and the denominator represent the same quantity in different units.

Dimensional Analysis: You might hear it called fence-posting. This is a tool that allows you to change from one unit to another. It's just a way of presenting stuff that makes dealing with the units easier. You deal with equivalent values all the time. Ever bought something with all quarters? You instinctively used dimensional analysis in your head.

Units: the good guys. Units are a measure of the magnitude of whatever you're talking about. They will help you identify the information you're given.

Steps to Solving Dimensional Analysis:
Step #1: write down the number and unit you would like to convert (starting point)
Step #2: write a multiplication sign followed by the conversion factor you wish to use.
What you go **from** should go on the bottom (in the denominator of the conversion factor). This is the unit you are converting out of.
What you go **to** should go on the top (in the numerator of the conversion factor). This is the unit you are converting to.
Step #3: repeat step 2 as necessary. It might take many conversions to get to the final unit.
Step #4: once you have obtained the unit you want, solve. Numbers on the top (in the numerator) are multiplied. Numbers on the bottom (in the denominator) are divided.

Onward to Stoichiometry …
Understanding Mole Ratios:

One of the first exercises I have my students do is write out every mole ratio of a balanced chemical reaction. This gets them used to the idea that any part of a reaction can be related to any other part. For convenience, I assign the reactants and products letters, so they're easier to talk about in class. Combustion reactions are convenient to use in stoichiometry problems because they give us interesting mole ratios. There could be a mole ratio of 1 to 1, but it's less interesting to speak about.

Example:
$C_3H_8 + 5\ O_2 \rightarrow 3\ CO_2 + 4\ H_2O$
A B C D
You can relate A to B, A to C, A to D, B to C, B to D, or C to D.

A to B: $\frac{\text{1 mole C3H8}}{\text{5 moles O2}}$	A to C: $\frac{\text{1 mole C3H8}}{\text{3 moles CO2}}$	A to D: $\frac{\text{1 mole C3H8}}{\text{4 moles H2O}}$
B to C: $\frac{\text{5 moles O2}}{\text{3 moles CO2}}$	B to D: $\frac{\text{5 moles O2}}{\text{4 moles H2O}}$	C to D: $\frac{\text{3 moles CO2}}{\text{4 moles H2O}}$

Note: The flipped versions of any of these are still valid mole ratios, but I think writing it out one way is good enough to get the point that they're related.

What is ideal stoichiometry? How does it differ from limiting reactant stoichiometry?
Ideal stoichiometry (as opposed to limiting reactant stoichiometry) assumes perfect reactions. In other words, no loss of reactants and 100% yield for products. Limiting reactant stoichiometry assumes more realistic conditions where only some reactant molecules go through the chemical reaction and give you product. Limiting reactant stoichiometry also acknowledges that you don't usually start with exactly enough of each reactant. Typically, something runs out first and limits the amount of product that can be formed.

Four Types of Ideal Stoichiometry Problems:
There are four types of problems you will likely be asked to solve for in ideal stoichiometry: mole to mole, mole to mass, mass to mole, and mass to mass. The steps to solving these problems are the same as any other conversion problem. For each reaction, you need a balanced chemical equation. Let's continue to use the combustion of propane.

Review of Using Conversion Factors to Convert Units:
Step #1: Start with given information
Step #2: Put units for the unit that needs to change on the bottom and units that you're changing to on the top
Step #3: Repeat step 2 until you get to the final unit
Step #4: Plug in the conversion factors
Step #5: Solve (top = multiply, bottom = divide)

Mole to Mole Problems: one step
Example 1: $C_3H_8 + 5\ O_2 \rightarrow 3\ CO_2 + 4\ H_2O$
Question: If 3.15 moles of O_2 react, how many moles of CO_2 could be formed?

Answer: 3.15 moles O_2 x $\frac{3\ moles\ CO2}{5\ moles\ O2}$ = 1.89 moles CO_2

Explanation: Start with your given amount of moles of O_2, then set up a conversion factor. The mole ratio between oxygen and CO_2 is $\frac{5\ moles\ O2}{3\ moles\ CO2}$ or $\frac{3\ moles\ CO2}{5\ moles\ O2}$. Since you need to cancel oxygen, you need the second version of the mole ratio with the five moles of O_2 on the bottom and the three moles of CO_2 on the top. This is the only conversion factor needed at this time. Solve by multiplying by three and dividing by five. Report your answers as your teacher requires. Most will say to go to the same sig figs as the given amount. 3.15 has three sig figs, so the answer would be 1.89 moles CO_2.

Mole to Mass Problems: two steps
Example 2: $C_3H_8 + 5\ O_2 \rightarrow 3\ CO_2 + 4\ H_2O$
Question: If 3.15 moles of O_2 react, what mass of CO_2 could be formed?

Answer: 3.15 moles O_2 x $\frac{3\ moles\ CO2}{5\ moles\ O2}$ x $\frac{44\ g\ CO2}{1\ mol\ CO2}$ = 83.16 g CO_2 or 83.2 g CO_2

Explanation: The problem starts out exactly the same as before by using the mole ratio of $\frac{3\ moles\ CO2}{5\ moles\ O2}$ to convert out of moles of O_2 and into moles of CO_2. Since the problem asks for the

grams of CO_2, you need the molar mass of carbon dioxide to convert out of moles of CO_2 and into grams of CO_2. Note that the molar mass of CO_2 could be written $\frac{1\ mol\ CO2}{44\ g\ CO2}$ or $\frac{44\ g\ CO2}{1\ mol\ CO2}$. Because you're already in moles of CO_2 by the end of the mole ratio step, you need the second version of the molar mass conversion factor with the grams on top. Solve as usual by multiplying across the top and dividing across the bottom.
Tip: It shouldn't matter which order you do the steps in mathematically as long as you do everything.

Mass to Mole Problems: two steps
Example 3: $C_3H_8 + 5\ O_2 \rightarrow 3\ CO_2 + 4\ H_2O$
Question: In an ideal reaction, how many moles of water (H_2O) could be formed from a reaction of 74.5 g O_2?

Answer: 74.5 g O_2 x $\frac{1\ mole\ O2}{32\ g\ O2}$ x $\frac{4\ moles\ H2O}{5\ moles\ O2}$ = 1.8625 moles H_2O or 1.86 moles H_2O

Explanation: Start with the given of 74.5 g O_2. The first conversion factor is a molar mass to get out of grams of O_2 and into moles of O_2 because you can't do a mole ratio until you're in the unit of moles. Once you're in moles of O_2, do another conversion from moles of O_2 to moles of H_2O. This step requires the mole ratio of oxygen to water. Of the two possible versions of the conversion factor, you need the one with the moles of water on top because that is the unit you were asked to switch to. Solve by multiplying across the top and dividing anything on the bottom. To three sig figs, the answer is 1.86 moles H_2O.

Mass to Mass Problems: three steps
Example 4: $C_3H_8 + 5\ O_2 \rightarrow 3\ CO_2 + 4\ H_2O$
Question: In an ideal reaction, what mass of water (H_2O) in grams could be formed from a reaction of 74.5 g O_2?

Answer: 74.5 g O_2 x $\frac{1\ mole\ O2}{32\ g\ O2}$ x $\frac{4\ moles\ H2O}{5\ moles\ O2}$ x $\frac{18\ g\ H2O}{1\ mol\ H2O}$ = 33.525 g H_2O or 33.5 g H_2O

Explanation: This problem is identical to the one before except for the tiny detail of the end unit. This time, you're starting out with 74.5 g O_2 and asked how many grams of water could be formed. The solution involves the exact same two steps in the previous example and one more conversion. Instead of ending with moles of water, use the molar mass of water to convert to grams. At the end of the last problem, you were in moles H_2O. To convert out of moles of water and into grams of water, you need to multiply by 18 g H_2O. The final answer to three sig figs is 33.5 g H_2O.

Note: Your teacher might be particular about how you show these conversion factors, or they may teach you another way entirely. I've seen stoichiometry done with ratios too. It works, but I think this method has a better chance of letting students see where they started and watch the progress through the problem until there's a solution. I require all my students to write a number, unit, and label on every step. It's somewhat tedious, but it's also the best way of tracking where you came from and where you're going.

Name: ______________________________ Date: ______________ Period: ______

Note: Some of you probably need the practice putting your name on stuff.

Worksheet 1: Mole to Mole Ideal Stoichiometry

$$2\ Na_2CO_3 \rightarrow 4\ Na + 2\ C + 3\ O_2$$

A B C D

1. What are the mol ratios that exist in this chemical formula? (The letters underneath the equation are arbitrary. I put them there to make it easier to explain the relationships.)

2. How many moles of Na_2CO_3 need to decompose to form 4.64 moles of O_2?

3. How many moles of Na could be formed if 3.5 moles of Na_2CO_3 decomposed?

4. How many moles of C can be formed if 0.4564 moles of Na are also formed?

5. How many moles of O_2 are formed if 32.8 moles of Na_2CO_3 are decomposed?

6. How many moles of O_2 would be formed if 0.7953 moles of Na were also formed?

7. How many moles of Na would be formed if 6.87 moles of C also formed?

Name: ________________________________ Date: ______________ Period: ______

Note: Some of you probably need the practice putting your name on stuff.

Worksheet 1 Key: Mole to Mole Ideal Stoichiometry

$$2\ Na_2CO_3 \rightarrow 4\ Na + 2\ C + 3\ O_2$$

A B C D

1. What are the mol ratios that exist in this chemical formula? (The letters underneath the equation are arbitrary. I put them there to make it easier to explain the relationships.)

A to B: $\frac{2\ mol\ \text{Na2CO3}}{4\ mol\ Na}$	A to C: $\frac{2\ \text{mol Na2CO3}}{2\ mol\ C}$	A to D: $\frac{2\ \text{mol Na2CO3}}{3\ mol\ O2}$
B to C: $\frac{4\ mol\ Na}{2\ mol\ C}$	B to D: $\frac{4\ mol\ Na}{3\ mol\ O2}$	C to D: $\frac{2\ mol\ C}{3\ mol\ O2}$

2. How many moles of Na_2CO_3 need to decompose to form 4.64 moles of O_2?

$4.64\ \text{mol}\ O_2 \times \frac{2\ mol\ \text{Na2CO3}}{3\ mol\ O2} = 3.09\ \text{mol}\ Na_2CO_3$

Explanation:

- Start with the given, which is 4.64 moles O_2.
- Set up a conversion factor with the mole ratio between Na_2CO_3 and O_2 ($\frac{2\ mol\ \text{Na2CO3}}{3\ mol\ O2}$). The numbers for the mole ratio come directly from the balanced chemical equation. The unit is moles. The label/chemical species we want to end up with is Na_2CO_3, which is why it's on top.
- Multiply values on the top of the conversion factor. Divide by values on the bottom of the conversion factor.
- Report answer: The answer to 3 significant figures should be 3.09 mol Na_2CO_3.

3. How many moles of Na could be formed if 3.5 moles of Na_2CO_3 decomposed?

$3.5\ \text{mol}\ Na_2CO_3 \times \frac{4\ mol\ Na}{2\ mol\ \text{Na2CO3}} = 7.0\ \text{mol Na}$

Explanation: Start with the given, which is 3.5 mol Na_2CO_3. Then, set up a conversion factor using the mole ratio between Na and Na_2CO_3 which is ($\frac{4\ mol\ Na}{2\ mol\ \text{Na2CO3}}$). The moles of Na are on top because that is what I'd like to go to. The moles of Na_2CO_3 are on the bottom because that is what I am seeking to cancel. When I do the mathematical operations, I get 7. To 2 significant figures and adding a unit and a chemical label, the answer is 7.0 mol Na.

$$2\ Na_2CO_3 \rightarrow 4\ Na + 2\ C + 3\ O_2$$

4. How many moles of C can be formed if 0.4564 moles of Na are also formed?

0.4564 mol of Na x $\frac{2\ mol\ C}{4\ mol\ Na}$ = 0.2282 mol C

Explanation:

- Write down the given 0.4564 mol of Na.
- Set up the mole ratio between C and Na, which is ($\frac{2\ mol\ C}{4\ mol\ Na}$).
- Multiply by 2 and divide by 4.
- Report the answer to 4 significant figures: 0.2282 mol C.

5. How many moles of O_2 are formed if 32.8 moles of Na_2CO_3 are decomposed?

32.8 mol Na_2CO_3 x $\frac{3\ mol\ O2}{2\ mol\ Na2CO3}$ = 49.2 mol O_2

Explanation: Start with the given of 32.8 mol Na_2CO_3. Set up the mole ratio with O_2 on top and the moles of Na_2CO_3 on the bottom. It should look like this: $\frac{3\ mol\ O2}{2\ mol\ Na2CO3}$. Next, solve. Multiply the numerical value of the given by 3 and divide by 2. The answer to 3 significant figures is 49.2 mol O_2.

6. How many moles of O_2 would be formed if 0.7953 moles of Na were also formed?

0.7953 mol Na x $\frac{3\ mol\ O2}{4\ mol\ Na}$ = 0.5965 mol O_2

Explanation:

- Write down the given information: 0.7953 mol Na.
- Set up the mole ratio between O_2 and Na. It should look like this: $\frac{3\ mol\ O2}{4\ mol\ Na}$ because you want the moles of O_2 on the top and the moles of Na on the bottom.
- Multiply by 3 (top value) and divide by 4 (bottom value). It doesn't matter which you put in the calculator first as long as you do both.
- The given was 4 significant figures, so the final answer is 0.5965 mol O_2.

7. How many moles of Na would be formed if 6.87 moles of C also formed?

6.87 mol C x $\frac{4\ mol\ Na}{2\ mol\ C}$ = 13.7 mol Na

Explanation: Start by writing down the given. Remember, the given includes a number, a unit, and the label or chemical species. The given is 6.87 mol C. Set up the conversion factor with 4 mols Na on top and 2 mols C on the bottom. It should look like this: $\frac{4\ mol\ Na}{2\ mol\ C}$. Solve by taking the given value of 6.87 mol C and multiplying by 4 and dividing by 2. Report the answer to 3 significant figures since that's how many are in the given value. The final answer should be 13.7 mol Na. It came to 13.74, but to 3 SF, the last digit is dropped.

Name: ______________________________ Date: ______________ Period: ______

Worksheet 2: Mole to Mass Ideal Stoichiometry

____ $FeCl_3$ + ____ $(NH_4)_2S$ → ____ Fe_2S_3 + ____ NH_4Cl

Note: I'm going to include the molar masses I'm using because every periodic table will be a little bit different. Even if you use a more complicated periodic table, the answers should be close.

$FeCl_3 = \frac{162.3\ g\ FeCl3}{1\ mol\ FeCl3}$	$(NH_4)_2S = \frac{68.1\ g\ (NH4)2S}{1\ mol\ (NH4)2S}$	$Fe_2S_3 = \frac{207.9\ g\ Fe2S3}{1\ mol\ Fe2S3}$	$NH_4Cl = \frac{53.5\ g\ NH4Cl}{1\ mol\ NH4Cl}$

1. If 34 moles of $FeCl_3$ reacted, how many grams of Fe_2S_3 could be formed?

2. How many grams of NH_4Cl can be formed if 3.25 moles of Fe_2S_3 are also formed?

3. How many grams of $FeCl_3$ are needed to fully react 2.00 moles of $(NH_4)_2S$?

4. How many grams of $(NH_4)_2S$ are needed to fully react 0.8791 moles of $FeCl_3$?

5. If 6.89 moles of $(NH_4)_2S$ reacted, how many grams of Fe_2S_3 could be formed?

6. If 98 moles of NH_4Cl were formed, how many g of $FeCl_3$ must have reacted?

Name: ______________________________ Date: _____________ Period: ______

Worksheet 2 Key: Mole to Mass Ideal Stoichiometry

$2\ FeCl_3 + 3\ (NH_4)_2S \rightarrow Fe_2S_3 + 6\ NH_4Cl$

Note: I'm going to include the molar masses I'm using because every periodic table will be a little bit different. Even if you use a more complicated periodic table, the answers should be close.

$FeCl_3 = \frac{162.3\ g\ FeCl3}{1\ mol\ FeCl3}$	$(NH_4)_2S = \frac{68.1\ g\ (NH4)2S}{1\ mol\ (NH4)2S}$	$Fe_2S_3 = \frac{207.9\ g\ Fe2S3}{1\ mol\ Fe2S3}$	$NH_4Cl = \frac{53.5\ g\ NH4Cl}{1\ mol\ NH4Cl}$

1. If 34 moles of $FeCl_3$ reacted, how many grams of Fe_2S_3 could be formed?

$34\ mol\ FeCl_3 \times \frac{1\ mol\ Fe2S3}{2\ mol\ FeCl3} \times \frac{207.9\ g\ Fe2S3}{1\ mol\ Fe2S3} = 3500\ g\ Fe_2S_3$

Explanation:

- Write down the given information: 34 moles of $FeCl_3$
- Use the mole ratio $(\frac{1\ mol\ Fe2S3}{2\ mol\ FeCl3})$ to convert out of moles of $FeCl_3$ and into the moles of Fe_2S_3.
- Use the molar mass of Fe_2S_3 to convert to the moles of Fe_2S_3: $(\frac{207.9\ g\ Fe2S3}{1\ mol\ Fe2S3})$
- Divide by 2 (bottom of a conversion factor)
- Multiply by 207.9 (top of a conversion factor)
- Adjust the answer to 2 significant figures: 3500 g Fe_2S_3

2. How many grams of NH_4Cl can be formed if 3.25 moles of Fe_2S_3 are also formed?

$3.25\ mol\ Fe_2S_3 \times \frac{6\ mol\ NH4Cl}{1\ mol\ Fe2S3} \times \frac{53.5\ g\ NH4Cl}{1\ mol\ NH4Cl} = 1040\ g\ NH_4Cl$

Explanation: Start by writing down the given information. This time, you are given 3.25 moles of Fe_2S_3. Your initial unit is moles and the initial label is Fe_2S_3, so the mole ratio you need is $\frac{6\ mol\ NH4Cl}{1\ mol\ Fe2S3}$. This will let you convert to moles of NH_4Cl. Next, you'll need to get out of moles of NH_4Cl and into grams of that same chemical. Solve by multiplying the given by the 6 mol and the 53.5 g. When adjusted to 3 significant figures, the answer is 1040 g NH_4Cl.

$$2\ FeCl_3 + 3\ (NH_4)_2S \rightarrow Fe_2S_3 + 6\ NH_4Cl$$

$FeCl_3 = \frac{162.3\ g\ FeCl3}{1\ mol\ FeCl3}$	$(NH_4)_2S = \frac{68.1\ g\ (NH4)2S}{1\ mol\ (NH4)2S}$	$Fe_2S_3 = \frac{207.9\ g\ Fe2S3}{1\ mol\ Fe2S3}$	$NH_4Cl = \frac{53.5\ g\ NH4Cl}{1\ mol\ NH4Cl}$

3. How many grams of $FeCl_3$ are needed to fully react 2.00 moles of $(NH_4)_2S$?

$$2.00\ mol\ (NH_4)_2S \times \frac{2\ mol\ FeCl3}{3\ mol\ (NH4)2S} \times \frac{162.3\ g\ FeCl3}{1\ mol\ FeCl3} = 216\ g\ FeCl_3$$

Explanation:

- Write down the given: 2.00 mols $(NH_4)_2S$.
- Use the mole ratio to convert out of $(NH_4)_2S$ and into $FeCl_3$ $(\frac{2\ mol\ FeCl3}{3\ mol\ (NH4)2S})$
- Use the molar mass of $FeCl_3$ $(\frac{162.3\ g\ FeCl3}{1\ mol\ FeCl3})$ to convert out of moles and into grams.
- Solve by multiplying by 2 and 162.3 (both are top values) and dividing by 3 (bottom of a conversion factor)
- Round the answer to 3 sig figs: 216 g $FeCl_3$

4. How many grams of $(NH_4)_2S$ are needed to fully react 0.8791 moles of $FeCl_3$?

$$.8791\ mol\ FeCl_3 \times \frac{3\ mol\ (NH4)2S}{2\ mol\ FeCl3} \times \frac{68.1\ g\ (NH4)2S}{1\ mol\ (NH4)2S} = 89.80\ g\ (NH_4)_2S$$

Explanation: Start by writing down the given, which is 0.8791 mol $FeCl_3$. Use a mole ratio to switch chemicals from $FeCl_3$ to $(NH_4)_2S$. It should look like this $\frac{3\ mol\ (NH4)2S}{2\ mol\ FeCl3}$, so you get out of the $FeCl_3$ and into the $(NH_4)_2S$. Once in moles, convert to grams of $(NH_4)_2S$ by using its molar mass. The conversion looks like this $\frac{68.1\ g\ (NH4)2S}{1\ mol\ (NH4)2S}$ so you can get to grams. Once everything on top has been multiplied and on the bottom has been divided, the answer is rounded to 4 significant figures. The answer is 89.80 g $(NH_4)_2S$.

5. If 6.89 moles of $(NH_4)_2S$ reacted, how many grams of Fe_2S_3 could be formed?

$$6.89\ mol\ (NH_4)_2S \times \frac{1\ mol\ Fe2S3}{3\ mol\ (NH4)2S} \times \frac{207.9\ g\ Fe2S3}{1\ mol\ Fe2S3} = 477\ g\ Fe_2S_3$$

Explanation:

- Write down the given: 6.89 mol $(NH_4)_2S$
- Switch the chemicals by using the mole ratio of $\frac{1\ mol\ Fe2S3}{3\ mol\ (NH4)2S}$
- Use the molar mass of Fe_2S_3 $(\frac{207.9\ g\ Fe2S3}{1\ mol\ Fe2S3})$ to change from moles to grams
- Multiply values that are on the top and divide things on the bottom
- Round the answer to 3 sig figs: 477 g Fe_2S_3

$$2\ FeCl_3 + 3\ (NH_4)_2S \rightarrow Fe_2S_3 + 6\ NH_4Cl$$

$FeCl_3 = \frac{162.3\ g\ FeCl3}{1\ mol\ FeCl3}$	$(NH_4)_2S = \frac{68.1\ g\ (NH4)2S}{1\ mol\ (NH4)2S}$	$Fe_2S_3 = \frac{207.9\ g\ Fe2S3}{1\ mol\ Fe2S3}$	$NH_4Cl = \frac{53.5\ g\ NH4Cl}{1\ mol\ NH4Cl}$

6. If 98 moles of NH_4Cl were formed, how many g of $FeCl_3$ must have reacted?

$98\ mol\ NH_4Cl \times \frac{2\ mol\ FeCl3}{6\ mol\ NH4Cl} \times \frac{162.3\ g\ FeCl3}{1\ mol\ FeCl3} = 5300\ g\ FeCl_3$

Explanation: Start by writing down the given of 98 mol NH_4Cl. The first thing that needs doing is switching chemicals, so use a mole ratio of $\frac{2\ mol\ FeCl3}{6\ mol\ NH4Cl}$ to accomplish this. When using conversion factor method, the units and labels you're canceling should be on the opposite side of the fraction. My original number started on the top, so to cancel it, I put that same unit on the bottom. After that, you need to switch between the moles of $FeCl_3$ to grams of the same thing. That requires a molar mass set up like this: $\frac{162.3\ g\ FeCl3}{1\ mol\ FeCl3}$. After multiplying top values and dividing bottom values, report the answer to 2 significant figures. The answer is 5300 g $FeCl_3$.

Name: ______________________________ Date: ______________ Period: ______

Worksheet 3: Mass to Mole Ideal Stoichiometry

____ FeS_2 + ____ O_2 → ____ Fe_2O_3 + ____ SO_2

$FeS_2 = \frac{120\ g\ FeS2}{1\ mol\ FeS2}$	$O_2 = \frac{32\ g\ O2}{1\ mol\ O2}$	$Fe_2O_3 = \frac{159.6\ g\ Fe2O3}{1\ mol\ Fe2O3}$	$SO_2 = \frac{64.1\ g\ SO2}{1\ mol\ SO2}$

1. If 68.3 g of FeS_2 reacted, how many moles of O_2 would also react?

2. If 23.6 g of Fe_2O_3 were formed, how many moles of SO_2 would also be formed?

3. If 2.51 g of O_2 reacted, how many moles of Fe_2O_3 would form?

4. If 5.00 g of SO_2 were formed, how many moles of FeS_2 reacted?

5. How many moles of Fe_2O_3 would form if 6.1 g O_2 reacted?

6. How many moles of O_2 are required to form 0.511 g SO_2?

Name: ______________________________ Date: ______________ Period: ______

Worksheet 3 Key: Mass to Mole Ideal Stoichiometry

$$4\ FeS_2 + 11\ O_2 \rightarrow 2\ Fe_2O_3 + 8\ SO_2$$

$FeS_2 = \frac{120\ g\ FeS2}{1\ mol\ FeS2}$	$O_2 = \frac{32\ g\ O2}{1\ mol\ O2}$	$Fe_2O_3 = \frac{159.6\ g\ Fe2O3}{1\ mol\ Fe2O3}$	$SO_2 = \frac{64.1\ g\ SO2}{1\ mol\ SO2}$

1. If 68.3 g of FeS_2 reacted, how many moles of O_2 would also react?

$68.3\ g\ FeS_2 \times \frac{1\ mol\ FeS2}{120\ g\ FeS2} \times \frac{11\ mol\ O2}{4\ mol\ FeS2} = 1.57\ mol\ O_2$

Explanation:

- Start with the given, which is 68.3 g of FeS_2
- Cancel g of FeS_2, by using the molar mass $(\frac{1\ mol\ FeS2}{120\ g\ FeS2})$
- Use a mole ratio to convert out of FeS_2 and into moles of O_2 $(\frac{11\ mol\ O2}{4\ mol\ FeS2})$. The numbers for the mole ratio come directly from the balanced chemical equation.
- Divide by the bottom values 120 g and 4 mol and multiply by the 11 mol
- Round the answer to 3 sig figs and add unit and species label: 1.57 mol O_2

2. If 23.6 g of Fe_2O_3 were formed, how many moles of SO_2 would also be formed?

$23.6\ g\ Fe_2O_3 \times \frac{1\ mol\ Fe2O3}{159.6\ g\ Fe2O3} \times \frac{8\ mol\ SO2}{2\ mol\ Fe2O3} = 0.591\ mol\ SO_2$

Explanation: Start with the given, which is 23.6 g of Fe_2O_3. Set up the first conversion factor to cancel the grams of Fe_2O_3 and go to the moles of Fe_2O_3. The conversion factor looks like this: $\frac{1\ mol\ Fe2O3}{159.6\ g\ Fe2O3}$. Next, use a mole ratio of $\frac{8\ mol\ SO2}{2\ mol\ Fe2O3}$ to convert from the moles of Fe_2O_3 to the moles of SO_2. Solve by multiplying across the top values and dividing the bottom values. The answer comes to 0.591 moles of SO_2.

Worksheet 3 Key continued: Mass to Mole Ideal Stoichiometry

$$4\ FeS_2 + 11\ O_2 \rightarrow 2\ Fe_2O_3 + 8\ SO_2$$

3. If 2.51 g of O_2 reacted, how many moles of Fe_2O_3 would form?

$$2.51\ g\ O_2 \times \frac{1\ mol\ O2}{32\ g\ O2} \times \frac{2\ mol\ Fe2O3}{11\ mol\ O2} = 0.0143\ mol\ Fe_2O_3$$

Explanation:

- Start with the given, which is 2.51 g of O_2.
- Use the molar mass of O_2 $(\frac{1\ mol\ O2}{32\ g\ O2})$ to convert from the grams to the moles of O_2.
- Next, use the mole ratio $(\frac{2\ mol\ Fe2O3}{11\ mol\ O2})$ to convert out of O_2 and to Fe_2O_3.
- Multiply by the 2 moles and divide by the 32 g and the 11 moles
- Report the answer to 3 significant figures: 0.0143 mol Fe_2O_3

4. If 5.00 g of SO_2 were formed, how many moles of FeS_2 reacted?

$$5.00\ g\ SO_2 \times \frac{1\ mol\ SO2}{64.1\ g\ SO2} \times \frac{4\ mol\ FeS2}{8\ mol\ SO2} = 0.0390\ mol\ FeS_2$$

Explanation: Start with the given, which is 5.00 g of SO_2. Use the molar mass of SO_2 to convert out of the grams and into the moles. To do so, you need to set it up this way $\frac{1\ mol\ SO2}{64.1\ g\ SO2}$ because the given is assumed to be the top part of a fraction here. That step will leave the answer in moles of SO_2, but the problem asks for the moles of FeS_2. That's where stoichiometry kicks in. Use a mole ratio of $\frac{4\ mol\ FeS2}{8\ mol\ SO2}$ to cancel the moles of SO_2 and end up in the moles of FeS_2. Once the problem is set up, you use a calculator to solve. It's usually best to start with the given. The numbers on the top get multiplied, the numbers on the bottom get divided by. The final answer needs to be reported to 3 significant figures here. The final answer is 0.0390 mols FeS_2.

5. How many moles of Fe_2O_3 would form if 6.1 g O_2 reacted?

$$6.1\ g\ O_2 \times \frac{1\ mol\ O2}{32\ g\ O2} \times \frac{2\ mol\ Fe2O3}{11\ mol\ O2} = 0.035\ mol\ Fe_2O_3$$

Explanation:

- Start by writing down the given information: 6.1 g O_2
- Use the molar mass of O_2 $(\frac{1\ mol\ O2}{32\ g\ O2})$ to change the units from grams to moles
- Use the mole ratio $(\frac{2\ mol\ Fe2O3}{11\ mol\ O2})$ to change the chemical species from O_2 to Fe_2O_3
- Solve by multiplying values that are on the top part of the fractions and dividing by the numbers on the bottom portion of the fractions
- The final answer to 2 sig figs is 0.035 moles Fe_2O_3

$$4\ FeS_2 + 11\ O_2 \rightarrow 2\ Fe_2O_3 + 8\ SO_2$$

6. How many moles of O_2 are required to form 0.511 g SO_2?

$$.511\ g\ SO_2 \times \frac{1\ mol\ SO2}{64.1\ g\ SO2} \times \frac{11\ mol\ O2}{8\ mol\ SO2} = 0.0110\ g\ SO_2$$

Explanation: Start with the given, which is 0.511 g SO_2. The current unit is grams and the current chemical label is SO_2. To cancel the g of SO_2, the molar mass of SO_2 will be set up as a conversion factor with grams on the bottom and moles of the same chemical up top. It will look like this: $\frac{1\ mol\ SO2}{64.1\ g\ SO2}$. Now, the current unit and label will be moles of SO_2. To change out of this and to moles of O_2, we use a mole ratio. The mole ratio looks like this: $\frac{11\ mol\ O2}{8\ mol\ SO2}$ because to cancel the SO_2 it needs to be opposite of where it currently is on top. Solve by dividing by 64.1 g SO_2, multiplying by 11 mol O_2, and dividing by 8 moles of SO_2. One only needs to plug the numbers into the calculator. The units and labels will sort themselves. The final answer is 0.0110 g SO_2. This one was tricky. The calculator answer says 0.01096139, but to get that to be 3 significant figures, we look to the 5th number in line because the first 0 doesn't count. A 6 rounds up. 9 also rounds up, making that 0 after the 1 also a 1, but then we still need a third significant figure, so we add a 0.

Name: ________________________________ Date: _______________ Period: ______

Worksheet 4: Mass to Mass Ideal Stoichiometry

____ $KClO_3$+ ____ P_4 → ____ P_4O_{10} + ____ KCl

$KClO_3 = \frac{122.6\ g\ \text{KClO3}}{1\ mol\ \text{KClO3}}$	$P_4 = \frac{124\ g\ \text{P4}}{1\ mol\ \text{P4}}$	$P_4O_{10} = \frac{284\ g\ \text{P4O10}}{1\ mol\ \text{P4O10}}$	$KCl = \frac{74.6\ g\ \text{KCl}}{1\ mol\ \text{KCl}}$

1. If 300. g of P_4 reacted, how many g of KCl would form?

2. If 60.5 g $KClO_3$ reacted, how many g of P_4 would also react?

3. If 25.1 g KCl formed, how many g of P_4O_{10} also formed?

4. How many g of P_4 would are required to form 0.69 g of P_4O_{10}?

5. What mass (in grams) of P_4O_{10} would be formed if 100.2 g KCl also formed?

6. What mass (in g) of $KClO_3$ are needed to form 5.00 g KCl?

Name: ______________________________ Date: ______________ Period: ______

Worksheet 4 Key: Mass to Mass Ideal Stoichiometry

$10\ KClO_3 + 3\ P_4 \rightarrow 3\ P_4O_{10} + 10\ KCl$

$KClO_3 = \frac{122.6\ g\ KClO3}{1\ mol\ KClO3}$	$P_4 = \frac{124\ g\ P4}{1\ mol\ P4}$	$P_4O_{10} = \frac{284\ g\ P4O10}{1\ mol\ P4O10}$	$KCl = \frac{74.6\ g\ KCl}{1\ mol\ KCl}$

1. If 300. g of P_4 reacted, how many g of KCl would form?

$$300.\ g\ P_4 \times \frac{1\ mol\ P4}{124\ g\ P4} \times \frac{10\ mol\ KCl}{3\ mol\ P4} \times \frac{74.6\ g\ KCl}{1\ mol\ KCl} = 602\ g\ KCl$$

Explanation: Mass to mass problems are the longest "normal" ideal stoichiometry problems. I guess you could make them longer by asking for a conversion to or from grams to kilograms, but I don't do that to my students because that's a lot of conversions.

- As usual, start with the given 300. g P_4
- Get to moles of the given by using the molar mass of the P_4 ($\frac{1\ mol\ P4}{124\ g\ P4}$)
- Use a mole ratio between the two chemicals you're working with ($\frac{10\ mol\ KCl}{3\ mol\ P4}$) to get out of P_4 and into KCl.
- Use the molar mass ($\frac{74.6\ g\ KCl}{1\ mol\ KCl}$) to cancel moles and get to grams of KCl
- Solve by dividing the given by values on the bottom and multiplying by numbers on the top
- Adjust to 3 significant figures and report the answer: 602 g KCl

2. If 60.5 g $KClO_3$ reacted, how many g of P_4 would also react?

$$60.5\ g\ KClO_3 \times \frac{1\ mol\ KClO3}{122.6\ g\ KClO3} \times \frac{3\ mol\ P4}{10\ mol\ KClO3} \times \frac{124\ g\ P4}{1\ mol\ P4} = 18.4\ g\ P_4$$

Explanation: Start with the given of 60.5 g $KClO_3$. Use the molar mass ($\frac{1\ mol\ KClO3}{122.6\ g\ KClO3}$) of $KClO_3$ to switch out of grams and into moles. Then, use the mole ratio ($\frac{3\ mol\ P4}{10\ mol\ KClO3}$) to convert out of $KClO_3$ and into 3 mol P_4. This will allow you to switch the chemicals to the one you were asked about. Next, use the molar mass of P_4 to get out of the unit of moles and into the unit of grams. It will look like this: ($\frac{124\ g\ P4}{1\ mol\ P4}$). Finally, solve. Start with the given. Numbers on the top of conversion factors are multiplied. Numbers on the bottom of conversion factors are divided. The final answer to 3 significant figures is 18.4 g P_4.

Worksheet 4 Key continued: Mass to Mass Ideal Stoichiometry

$10\ KClO_3 + 3\ P_4 \rightarrow 3\ P_4O_{10} + 10\ KCl$

3. If 25.1 g KCl formed, how many g of P_4O_{10} also formed?

$$25.1 \text{ g KCl} \times \frac{1\ mol\ KCl}{74.6\ g\ KCl} \times \frac{3\ mol\ P4O10}{10\ mol\ KCl} \times \frac{284\ g\ P4O10}{1\ mol\ P4O10} = 28.7 \text{ g } P_4O_{10}$$

Explanation:

- Start with the given: 25.1 g KCl
- Use the molar mass of KCl ($\frac{1\ mol\ KCl}{74.6\ g\ KCl}$) to switch out of grams and into moles
- After that, use the mole ratio between KCl and P_4O_{10} ($\frac{3\ mol\ P4O10}{10\ mol\ KCl}$) to switch chemicals
- Use the molar mass of P_4O_{10} ($\frac{284\ g\ P4O10}{1\ mol\ P4O10}$) to get out of moles of P_4O_{10} and go to the grams of P_4O_{10}
- Solve by multiplying the given across the top and dividing across the bottom.
- The final answer will be rounded to 3 significant figures: 28.7 g P_4O_{10}.

4. How many g of P_4 would are required to form 0.69 g of P_4O_{10}?

$$.69 \text{ g } P_4O_{10} \times \frac{1\ mol\ P4O10}{284\ g\ P4O10} \times \frac{3\ mol\ P4}{3\ mol\ P4O10} \times \frac{124\ P4}{1\ mol\ P4} = 0.30 \text{ g } P_4$$

Explanation: Start with the given amount of P_4O_{10} (.69 g P_4O_{10}). The first thing that needs doing is to get to moles of the given. To do this, divide by the molar mass of P_4O_{10}. The conversion factor will look like this: $\frac{1\ mol\ P4O10}{284\ g\ P4O10}$. The mole ratio ($\frac{3\ mol\ P4}{3\ mol\ P4O10}$) will let you switch the chemicals between P_4O_{10} and P_4. Then, the molar mass of P_4 ($\frac{124\ P4}{1\ mol\ P4}$) will be used to change the units from moles to grams. Solve by multiplying the given by the numbers on the top of the conversion factors and dividing by the numbers on the bottom of the conversion factors. You don't technically have to multiply and divide by 3 as they will cancel each other out, but do not reduce the mole ratio. The answer will be rounded to 2 significant figures this time because the given is only 2 sig figs. The answer is 0.30 g P_4.

Worksheet 4 Key continued: Mass to Mass Ideal Stoichiometry

$$10\ KClO_3 + 3\ P_4 \rightarrow 3\ P_4O_{10} + 10\ KCl$$

5. What mass (in grams) of P_4O_{10} would be formed if 100.2 g KCl also formed?

100.2 g KCl x $\frac{1\ mol\ KCl}{74.6\ g\ KCl}$ x $\frac{3\ mol\ P4O10}{10\ mol\ KCl}$ x $\frac{284\ g\ P4O10}{1\ mol\ P4O10}$ = 114.4 g P_4O_{10}

Explanation:

- Write the given: 100.2 g KCl
- Use the molar mass of KCl ($\frac{1\ mol\ KCl}{74.6\ g\ KCl}$) to obtain moles of the given.
- Use the molar mass to convert from KCl to P_4O_{10} ($\frac{3\ mol\ P4O10}{10\ mol\ KCl}$).
- Switch the units from moles to grams by using the molar mass of P_4O_{10} ($\frac{284\ g\ P4O10}{1\ mol\ P4O10}$)
- Solve by multiplying the given by all numbers on the tops of the conversions and dividing by numbers on the bottom.
- The answer will be rounded to 4 significant figures: 114.4 g P_4O_{10}

6. What mass (in g) of $KClO_3$ are needed to form 5.00 g KCl?

5.00 g KCl x $\frac{1\ mol\ KCl}{74.6\ g\ KCl}$ x $\frac{10\ mol\ KClO3}{10\ mol\ KCl}$ x $\frac{122.6\ g\ KClO3}{1\ mol\ KClO3}$ = 8.22 g $KClO_3$

Explanation: Start with the given amount KCl (5.00 g KCl). Get out of grams and into moles of KCl by using the molar mass set up with the grams on the bottom ($\frac{1\ mol\ KCl}{74.6\ g\ KCl}$). The molar mass will then be used to convert from KCl to $KClO_3$ ($\frac{10\ mol\ KClO3}{10\ mol\ KCl}$). This is a necessary step even though the values of the mole ratio will cancel each other. It still lets the chemical labels be switched. The last conversion is also a molar mass, but the mass portion is on top ($\frac{122.6\ g\ KClO3}{1\ mol\ KClO3}$). Solve by starting with the given and multiplying by top values and dividing by bottom values of the conversion factors. The answer should be 8.22 g $KClO_3$ because you need to round to 3 significant figures.

Cheat Sheet of Ideal Stoichiometry Facts:

The next worksheets will ask you to predict the "type" of ideal stoichiometry conversion you'll be doing. "Type" refers to the unit are you starting with (given) to unit of the new chemical.

All "givens" will have a number, a unit, and a label.

Two Types of Conversion Factors:

1. Mole ratios allow you to switch between chemicals
2. Molar masses allow you to switch between grams and moles

Mole to Mole: 1 conversion step

$$\text{given in mol x } \underbrace{\frac{\textit{moles of chemical you are going to}}{\text{mol of the given}}}_{\text{mole ratio}} = \text{answer}$$

Mole to Mass: 2 conversion steps

$$\text{given in mol x } \underbrace{\frac{\textit{moles of chemical you are going to}}{\text{mol of the given}}}_{\text{mole ratio}} \text{ x } \underbrace{\frac{\text{mass of the new chemical}}{\text{1 mol of the new chemical}}}_{\text{molar mass}} = \text{answer}$$

Mass to Mole: 2 conversion steps

$$\text{given in g x } \underbrace{\frac{\text{1 mol of the given}}{\text{mass of the given}}}_{\text{molar mass}} \text{ x } \underbrace{\frac{\text{moles of new chemical}}{\text{mol of the given}}}_{\text{mol ratio}} = \text{answer}$$

Mass to mass: 3 conversion steps

$$\text{given in g x } \underbrace{\frac{\text{1 mol of the given}}{\text{mass of the given}}}_{\text{molar mass}} \text{ x } \underbrace{\frac{\text{moles of new chemical}}{\text{mol of the given}}}_{\text{mol ratio}} \text{ x } \underbrace{\frac{\textit{mass of new chemical}}{1 \textit{ mol new chemical}}}_{\text{molar mass}} =$$

Name: ________________________________ Date: ______________ Period: ______

Worksheet 5: Ideal Stoichiometry Mixed Self-Quiz 1

$$S_8 + 12\ O_2 \rightarrow 8\ SO_3$$

Directions: Answer each question using the conversion factor method. Watch sig figs.

$S_8 = \frac{256.8\ g\ S8}{1\ mol\ S8}$	$O_2 = \frac{32\ g\ O2}{1\ mol\ O2}$	$SO_3 = \frac{80.1\ g\ SO3}{1\ mol\ SO3}$

1. How many moles of S_8 are needed to produce 631 mol SO_3?
Type: # of conversion factors:

2. How many grams of O_2 would be required to fully react 2.3 mol S_8?
Type: # of conversion factors:

3. If 21.0 g SO_3 was formed, what mass (in g) of S_8 reacted?
Type: # of conversion factors:

4. How many mol of SO_3 would be formed if 6.11 mol O_2 reacted?
Type: # of conversion factors:

5. How many mol of O_2 would be required to react 1.94×10^{32} g S_8?
Type: # of conversion factors:

6. What mass (in g) of SO_3 could be formed from 351 g of O_2?
Type: # of conversion factors:

Name: ______________________________ Date: ______________ Period: ______

Worksheet 5 Key: Ideal Stoichiometry Mixed Self-Quiz 1

$$S_8 + 12\ O_2 \rightarrow 8\ SO_3$$

Directions: Answer each question using the conversion factor method. Watch sig figs.

$S_8 = \frac{256.8\ g\ S8}{1\ mol\ S8}$	$O_2 = \frac{32\ g\ O2}{1\ mol\ O2}$	$SO_3 = \frac{80.1\ g\ SO3}{1\ mol\ SO3}$

1. How many moles of S_8 are needed to produce 631 mol SO_3?

Type: mol to mol # of conversion factors: 1

$$631\ mol\ SO_3 \times \frac{1\ mol\ S8}{8\ mol\ SO3} = 78.9\ mol\ S_8$$

2. How many grams of O_2 would be required to fully react 2.3 mol S_8?

Type: mol to mass # of conversion factors: 2

$$2.3\ mol\ S_8 \times \frac{12\ mol\ O2}{1\ mol\ S8} \times \frac{32\ g\ O2}{1\ mol\ O2} = 880\ g\ O_2$$

3. If 21.0 g SO_3 was formed, what mass (in g) of S_8 reacted?

Type: mass to mass # of conversion factors: 3

$$21.0\ g\ SO_3 \times \frac{1\ mol\ SO3}{80.1\ g\ SO3} \times \frac{1\ mol\ S8}{8\ mol\ SO3} \times \frac{256.8\ g\ S8}{1\ mol\ S8} = 8.42\ g\ S_8$$

4. How many mol of SO_3 would be formed if 6.11 mol O_2 reacted?

Type: mol to mol # of conversion factors: 1

$$6.11\ mol\ O_2 \times \frac{8\ mol\ SO3}{12\ mol\ O2} = 4.07\ mol\ SO_3$$

5. How many mol of O_2 would be required to react 1.94×10^{32} g S_8?

Type: mass to mol # of conversion factors: 2

$$1.94 \times 10^{32}\ g\ S_8 \times \frac{1\ mol\ S8}{256.8\ g\ S8} \times \frac{12\ mol\ O2}{1\ mol\ S8} = 9.07 \times 10^{30}\ mol\ O_2$$

6. What mass (in g) of SO_3 could be formed from 351 g of O_2?

Type: mass to mass # of conversion factors: 3

$$351\ g\ of\ O_2 \times \frac{1\ mol\ O2}{32\ g\ O2} \times \frac{8\ mol\ SO3}{12\ mol\ O2} \times \frac{80.1\ g\ SO3}{1\ mol\ SO3} = 586\ g\ SO_3$$

Name: ________________________________ Date: ______________ Period: ______

Worksheet 6: Ideal Stoichiometry Mixed Self-Quiz 2

____ $AlBr_3$ + ____ K_2SO_4 → ____ KBr + ____ $Al_2(SO_4)_3$

Directions: Answer each question using the conversion factor method. Watch sig figs.

$AlBr_3 = \frac{266.7\ g\ AlBr3}{1\ mol\ AlBr3}$	$K_2SO_4 = \frac{174.3\ g\ K2SO4}{1\ mol\ K2SO4}$	$KBr = \frac{119\ g\ KBr}{1\ mol\ KBr}$	$Al_2(SO_4)_3 = \frac{342.3\ g\ Al2(SO4)3}{1\ mol\ Al2(SO4)3}$

1. How many g of $AlBr_3$ are required to fully react 2.61 g K_2SO_4?
Type: # of conversion factors:

2. How many moles of KBr can be formed if 1.25 mol of $Al_2(SO_4)_3$ are also formed?
Type: # of conversion factors:

3. How many moles of K_2SO_4 are needed to form 200. g KBr?
Type: # of conversion factors:

4. How many g of can $Al_2(SO_4)_3$ be formed if 3.62 g KBr are also formed?
Type: # of conversion factors:

5. What mass of K_2SO_4 (in g) are required to form 0.31 mol $Al_2(SO_4)_3$?
Type: # of conversion factors:

6. How many mol of KBr can be formed from 8.50 mol $AlBr_3$?
Type: # of conversion factors:

Name: ______________________________ Date: ______________ Period: ______

Worksheet 6 Key: Ideal Stoichiometry Mixed Self-Quiz 2

$$2\ AlBr_3 + 3\ K_2SO_4 \rightarrow 6\ KBr + Al_2(SO_4)_3$$

Directions: Answer each question using the conversion factor method. Watch sig figs.

$AlBr_3 = \frac{266.7\ g\ AlBr3}{1\ mol\ AlBr3}$	$K_2SO_4 = \frac{174.3\ g\ K2SO4}{1\ mol\ K2SO4}$	$KBr = \frac{119\ g\ KBr}{1\ mol\ KBr}$	$Al_2(SO_4)_3 = \frac{342.3\ g\ Al2(SO4)3}{1\ mol\ Al2(SO4)3}$

1. How many g of $AlBr_3$ are required to fully react 2.61 g K_2SO_4?

Type: mass to mass # of conversion factors: 3

$$2.61\ g\ K_2SO_4 \times \frac{1\ mol\ K2SO4}{174.3\ g\ K2SO4} \times \frac{2\ mol\ AlBr3}{3\ mol\ K2SO4} \times \frac{266.7\ g\ AlBr3}{1\ mol\ AlBr3} = 2.66\ g\ AlBr_3$$

2. How many moles of KBr can be formed if 1.25 mol of $Al_2(SO_4)_3$ are also formed?

Type: mol to mol # of conversion factors: 1

$$1.25\ mol\ Al_2(SO_4)_3 \times \frac{6\ mol\ KBr}{1\ mol\ Al2(SO4)3} = 7.50\ mol\ KBr$$

3. How many moles of K_2SO_4 are needed to form 200. g KBr?

Type: mass to mol # of conversion factors: 2

$$200.\ g\ KBr \times \frac{1\ mol\ KBr}{119\ g\ KBr} \times \frac{3\ mol\ K2SO4}{6\ mol\ KBr} = 0.840\ mol\ K_2SO_4$$

4. How many g of can $Al_2(SO_4)_3$ be formed if 3.62 g KBr are also formed?

Type: mass to mass # of conversion factors: 3

$$3.62\ g\ KBr \times \frac{1\ mol\ KBr}{119\ g\ KBr} \times \frac{1\ mol\ Al2(SO4)3}{6\ mol\ KBr} \times \frac{342.3\ g\ Al2(SO4)3}{1\ mol\ Al2(SO4)3} = 1.74\ g\ Al_2(SO_4)_3$$

5. What mass of K_2SO_4 (in g) are required to form 0.31 mol $Al_2(SO_4)_3$?

Type: mol to mass # of conversion factors: 2

$$.31\ mol\ Al_2(SO_4)_3 \times \frac{3\ mol\ K2SO4}{1\ mol\ Al2(SO4)3} \times \frac{174.3\ g\ K2SO4}{1\ mol\ K2SO4} = 160\ g\ K_2SO_4$$

6. How many mol of KBr can be formed from 8.50 mol $AlBr_3$?

Type: mol to mol # of conversion factors: 1

$$8.50\ mol\ AlBr_3 \times \frac{6\ mol\ KBr}{2\ mol\ AlBr3} = 25.5\ mol\ KBr$$

Name: ______________________________ Date: ______________ Period: ______

Worksheet 7: Ideal Stoichiometry Mixed Self-Quiz 3

____ V_2O_5 + ____ HCl → ____ $VOCl_3$ + ____ H_2O

Directions: Answer each question using the conversion factor method. Watch sig figs.

$V_2O_5 = \frac{181.8\ g\ V2O5}{1\ mol\ V2O5}$	$HCl = \frac{36.5\ g\ HCl}{1\ mol\ HCl}$	$VOCl_3 = \frac{173.4\ g\ VOCl3}{1\ mol\ VOCl3}$	$H_2O = \frac{18\ g\ H2O}{1\ mol\ H2O}$

1. What mass of V_2O_5 is needed to react 8.22 mol HCl?
Type: # of conversion factors:

2. How many mol of $VOCl_3$ can be produced from 200. mol HCl?
Type: # of conversion factors:

3. How many moles of H_2O can be formed from 9.33 g V_2O_5?
Type: # of conversion factors:

4. How many g of $VOCl_3$ will be produced if 0.648 mol H_2O are also produced?
Type: # of conversion factors:

5. What mass of V_2O_5 is required for the full reaction of 825 g HCl?
Type: # of conversion factors:

6. How many g of H_2O can be formed from 9.1×10^{-3} mol HCl?
Type: # of conversion factors:

Name: ________________________________ Date: ______________ Period: ______

Worksheet 7 Key: Ideal Stoichiometry Mixed Self-Quiz 3

$V_2O_5 + 6\ HCl \rightarrow 2\ VOCl_3 + 3\ H_2O$

Directions: Answer each question using the conversion factor method. Watch sig figs.

$V_2O_5 = \frac{181.8\ g\ V2O5}{1\ mol\ V2O5}$	$HCl = \frac{36.5\ g\ HCl}{1\ mol\ HCl}$	$VOCl_3 = \frac{173.4\ g\ VOCl3}{1\ mol\ VOCl3}$	$H_2O = \frac{18\ g\ H2O}{1\ mol\ H2O}$

1. What mass of V_2O_5 is needed to react 8.22 mol HCl?

Type: mol to mass # of conversion factors: 2

$8.22\ mol\ HCl \times \frac{1\ mol\ V2O5}{6\ mol\ HCl} \times \frac{181.8\ g\ V2O5}{1\ mol\ V2O5} = 249\ g\ V_2O_5$

2. How many mol of $VOCl_3$ can be produced from 200. mol HCl?

Type: mol to mol # of conversion factors: 1

$200.\ mol\ HCl \times \frac{2\ mol\ VOCl3}{6\ mol\ HCl} = 66.7\ mol\ VOCl_3$

3. How many moles of H_2O can be formed from 9.33 g V_2O_5?

Type: mass to mol # of conversion factors: 2

$9.33\ g\ V_2O_5 \times \frac{1\ mol\ V2O5}{181.8\ g\ V2O5} \times \frac{3\ mol\ H2O}{1\ mol\ V2O5} = 0.154\ mol\ H_2O$

4. How many g of $VOCl_3$ will be produced if 0.648 mol H_2O are also produced?

Type: mol to mass # of conversion factors: 2

$.648\ mol\ H_2O \times \frac{2\ mol\ VOCl3}{3\ mol\ H2O} \times \frac{173.4\ g\ VOCl3}{1\ mol\ VOCl3} = 74.9\ g\ VOCl_3$

5. What mass of V_2O_5 is required for the full reaction of 825 g HCl?

Type: mass to mass # of conversion factors: 3

$825\ g\ HCl \times \frac{1\ mol\ HCl}{36.5\ g\ HCl} \times \frac{1\ mol\ V2O5}{6\ mol\ HCl} \times \frac{181.8\ g\ V2O5}{1\ mol\ V2O5} = 685\ g\ V_2O_5$

6. How many g of H_2O can be formed from 9.1×10^{-3} mol HCl?

Type: mol to mass # of conversion factors: 2

$9.1 \times 10^{-3}\ mol\ HCl \times \frac{3\ mol\ H2O}{6\ mol\ HCl} \times \frac{18\ g\ H2O}{1\ mol\ H2O} = 8.2 \times 10^{-2}\ g\ H_2O$

Name: ________________________________ Date: ______________ Period: ______

Worksheet 8: Ideal Stoichiometry Mixed Self-Quiz 4

____ N_2 + ____ O_2 + ____ H_2O → ____ HNO_3

Directions: Answer each question using the conversion factor method. Watch sig figs.

$N_2 = \frac{28\ g\ N2}{1\ mol\ N2}$	$O_2 = \frac{32\ g\ O2}{1\ mol\ O2}$	$H_2O = \frac{18\ g\ H2O}{1\ mol\ H2O}$	$HNO_3 = \frac{63\ g\ HNO3}{1\ mol\ HNO3}$

1. How many moles of O_2 are required to fully react 0.236 mol H_2O?
Type: # of conversion factors:

2. How many g of HNO_3 can be formed from 6.36 mol N_2?
Type: # of conversion factors:

3. How many g of N_2 are needed to fully react 631 g H_2O?
Type: # of conversion factors:

4. How many moles of H_2O are needed to form 6.9 g HNO_3?
Type: # of conversion factors:

5. What mass of HNO_3 can be produced if 5.0×10^3 g of O_2 reacts?
Type: # of conversion factors:

6. How many moles of H_2O are needed to react 20.05 g of O_2?
Type: # of conversion factors:

Name: ______________________________ Date: ______________ Period: ______

Worksheet 8 Key: Ideal Stoichiometry Mixed Self-Quiz 4

$$2\ N_2 + 5\ O_2 + 2\ H_2O \rightarrow 4\ HNO_3$$

Directions: Answer each question using the conversion factor method. Watch sig figs.

$N_2 = \frac{28\ g\ N2}{1\ mol\ N2}$	$O_2 = \frac{32\ g\ O2}{1\ mol\ O2}$	$H_2O = \frac{18\ g\ H2O}{1\ mol\ H2O}$	$HNO_3 = \frac{63\ g\ HNO3}{1\ mol\ HNO3}$

1. How many moles of O_2 are required to fully react 0.236 mol H_2O?

Type: mol to mol # of conversion factors: 1

$.236\ mol\ H_2O \times \frac{5\ mol\ O2}{2\ mol\ H2O} = 0.590\ mol\ O_2$

2. How many g of HNO_3 can be formed from 6.36 mol N_2?

Type: mol to mass # of conversion factors: 2

$6.36\ mol\ N_2 \times \frac{4\ mol\ HNO3}{2\ mol\ N2} \times \frac{63\ g\ HNO3}{1\ mol\ HNO3} = 801\ g\ HNO_3$

3. How many g of N_2 are needed to fully react 631 g H_2O?

Type: mass to mass # of conversion factors: 3

$631\ g\ H_2O \times \frac{1\ mol\ H2O}{18\ g\ H2O} \times \frac{2\ mol\ N2}{2\ mol\ H2O} \times \frac{28\ g\ N2}{1\ mol\ N2} = 982\ g\ H_2O$

4. How many moles of H_2O are needed to form 6.9 g HNO_3?

Type: mass to mol # of conversion factors: 2

$6.9\ g\ HNO_3 \times \frac{1\ mol\ HNO3}{63\ g\ HNO3} \times \frac{2\ mol\ H2O}{4\ mol\ HNO3} = 0.055\ mol\ H_2O$

5. What mass of HNO_3 can be produced if 5.0×10^3 g of O_2 reacts?

Type: mass to mass # of conversion factors: 3

$5.0 \times 10^3\ g\ O_2 \times \frac{1\ mol\ O2}{32\ g\ O2} \times \frac{4\ mol\ HNO3}{5\ mol\ O2} \times \frac{63\ g\ HNO3}{1\ mol\ HNO3} = 7.9 \times 10^3\ g\ HNO_3$

6. How many moles of H_2O are needed to react 20.05 g of O_2?

Type: mass to mol # of conversion factors: 2

$20.05\ g\ O_2 \times \frac{1\ mol\ O2}{32\ g\ O2} \times \frac{2\ mol\ H2O}{5\ mol\ O2} = 0.2506\ mol\ H_2O$

*****END IDEAL STOICHIOMETRY*****

Limiting Reactant Stoichiometry:

What if something runs out first?

Limiting Reactant – the reactant that is completely used up first
Excess Reactant – the reactant that has some left over after the other is used up

In most chemical reactions, something is going to be consumed first. This is the limiting reactant. It's the reactant that determines how much of each product should be created.

Note: You may see the term "limiting reagent." These terms are used interchangeably, but there are slight differences in the definitions. Typically, you're safe enough if you stick with LR, which covers both.

When going through this part of the unit, I ask five questions for every problem set.
1. What is the limiting reactant (LR)?
2. How many grams of the excess reactant are left over?
3. How many grams of each product could be produced in an ideal reaction?
4. If ___ g of (insert name of one product) were actually produced, what is the percent yield?
5. Assuming the same percent yield, how much of the other product would actually be produced?
(**Note:** This can change to I give you a percent yield to use instead of use the answer from 4.)

Example: $C_3H_8 + 5\ O_2 \rightarrow 3\ CO_2 + 4\ H_2O$
Question: If 135 g of each reactant were present …
1. What is the limiting reactant (LR)?

Two Methods for Finding Limiting Reactants:

Method #1: Choose one of the reactants, assume it's the LR, change to the other reactant, and evaluate. If you have more than you need, you guessed the LR correctly. If you need more than you have, you guessed incorrectly, the other reactant is the LR.

$$135\text{ g } C_3H_8 \times \frac{1\ mol\ C3H8}{44\ g\ C3H8} \times \frac{5\ mol\ O2}{1\ mol\ C3H8} \times \frac{32\ g\ O2}{1\ mol\ O2} = 491\text{ g } O_2$$

Analysis and conclusion: The key question is: do I have what I need or do I need more than I have? I've made an assumption that C_3H_8 is the limiting reactant and asked how many grams of O_2 I need to react at the same time. I was given 135 g of each reactant, but in order for C_3H_8 to be the LR, I need 491 g of O_2. Therefore, I guessed wrong. The LR for this reaction must be O_2.

Note: It's a complete, annoying coincidence that the molar mass of C_3H_8 and CO_2 are the same.

Method #2: Start with both reactants, assume both are LR, go to the SAME product, and evaluate.

$$135 \text{ g } C_3H_8 \text{ x } \frac{1 \, mol \, C3H8}{44 \, g \, C3H8} \text{ x } \frac{4 \, mol \, H2O}{1 \, mol \, C3H8} \text{ x } \frac{18 \, g \, H2O}{1 \, mol \, H2O} = 221 \text{ g } H_2O$$

$$135 \text{ g } O_2 \text{ x } \frac{1 \, mol \, O2}{32 \, g \, O2} \text{ x } \frac{4 \, mol \, H2O}{5 \, mol \, O2} \text{ x } \frac{18 \, g \, H2O}{1 \, mol \, H2O} = 60.8 \text{ g } H_2O$$

Analysis and conclusion: The reactant that gives less of the chosen product is the LR. In this example, the O_2 resulted in fewer grams of water, therefore it must be the LR.

Caution: Do not go to different products! Your goal is to compare the amounts of the products formed. This does not work if you go to different products.

Note: You can stop after moles in order to answer the question. I tend to go that extra step because I know the third question is going to be mass of products anyway. The set of conversions that involves the LR is now half of that answer.

Commentary: If given a choice, I go with Method 2 because I'm more comfortable with that evaluation. However, Method 2 has a very common pitfall: people often circle the lower amount of product and call it the LR. The LR is a reactant; therefore, it can never be a product. It's the reactant that resulted in the smaller amount of product. Method 1 also works just fine, but when you guess incorrectly, the analysis can look confusing.

$C_3H_8 + 5\ O_2 \rightarrow 3\ CO_2 + 4\ H_2O$

2. How many grams of the excess reactant are left over? (Start with the LR and change to the other reactant. Then, subtract this amount from the available starting amount of that other reactant.)

$$135 \text{ g } O_2 \text{ x } \frac{1 \, mol \, O2}{32 \, g \, O2} \text{ x } \frac{1 \, mol \, C3H8}{5 \, mol \, O2} \text{ x } \frac{44 \, g \, C3H8}{1 \, mol \, C3H8} = 37.1 \text{ g } C_3H_8 \text{ used in the reaction}$$

Available – used = excess left over

$135 \text{ g } C_3H_8 - 37.1 \text{ g } C_3H_8 = 97.9 \text{ g } C_3H_8$ left over

Explanation: Start with the limiting reactant and change over to non-limiting reactant. This tells you how many grams of the nonLR you used in the chemical reaction. A lot of students stop there, but there is one more step. Subtract the answer you just got from the amount of starting non-limiting reactant you were given to find the amount of the excess reactant left over.

$C_3H_8 + 5\ O_2 \rightarrow 3\ CO_2 + 4\ H_2O$

3. How many grams of each product could be produced in an ideal reaction?

$$135\text{ g } O_2 \text{ x } \frac{1\ mol\ O2}{32\ g\ O2} \text{ x } \frac{3\ mol\ CO2}{5\ mol\ O2} \text{ x } \frac{18\ g\ CO2}{1\ mol\ CO2} = 45.6\text{ g } CO_2$$

$$135\text{ g } O_2 \text{ x } \frac{1\ mol\ O2}{32\ g\ O2} \text{ x } \frac{4\ mol\ H2O}{5\ mol\ O2} \text{ x } \frac{18\ g\ H2O}{1\ mol\ H2O} = 60.8\text{ g } H_2O$$

Explanation: Start with the LR and change over to each of the products in turn. These are your theoretical yields. That means you assume 100% of the limiting reactant results in product. It's the maximum amount you could produce.

4. If 38.4 g of CO_2 were actually produced, what is the percent yield? (answer to two decimals)

$$\text{Percent yield} = \frac{actual\ yield}{theoretical\ yield} \text{ x } 100 \qquad \text{Percent yield} = \frac{38.4\ g\ CO2}{45.6\ g\ CO2} \text{ x } 100 = 84.21\%$$

Explanation: Any percent is a part of something over the whole multiplied by 100. In this question, the grams of the CO_2 are the "actual yield." The theoretical yield comes from the previous question. Multiplying by 100 simply turns the number into a percent.

Caution: It's easy to flip the numbers, but always evaluate your answer. Ask yourself: could this happen? The answer cannot be above 100% and probably shouldn't be something crazy small either. (That last one might be a personal bias, but I tell my students, when I make up the questions, I typically don't give things less than 50% yield. If you're getting 0.22%, it's probably wrong.)

5. Assuming the same percent yield, how much of the other product would actually be produced?

$$\text{Actual yield} = \frac{percent\ yield\ x\ theoretical\ yield}{100} \qquad \text{Actual yield} = \frac{60.8\text{ g H2O } x\ 84.21\%}{100} = 51.2\text{ g } H_2O$$

Explanation: Turn the percent yield back into a decimal by dividing by 100 and multiply by the other actual yield.

Directions: Calculate the molar masses. Then, balance the equation and solve the problems.

$$4\ FeS_2 + 11\ O_2 \rightarrow 2\ Fe_2O_3 + 8\ SO_2$$

$FeS_2 = \frac{120\ g\ FeS2}{1\ mol\ FeS2}$	$O_2 = \frac{32\ g\ O2}{1\ mol\ O2}$	$Fe_2O_3 = \frac{159.6\ g\ Fe2O3}{1\ mol\ Fe2O3}$	$SO_2 = \frac{64.1\ g\ SO2}{1\ mol\ SO2}$

If given 36.0 g of FeS_2 and 52.2 g of O_2 ...

7. What is the limiting reactant?

$$36.0\ g\ FeS_2 \times \frac{1\ mol\ FeS}{120\ g\ Fes} \times \frac{2\ mol\ Fe2O3}{4\ mol\ FeS} \times \frac{159.6\ g\ Fe2O3}{1\ mol\ Fe2O3} = 23.9\ g\ Fe_2O_3$$

$$52.2\ g\ O_2 \times \frac{1\ mol\ O2}{32\ g\ O2} \times \frac{2\ mol\ Fe2O3}{11\ mol\ O2} \times \frac{159.6\ g\ Fe2O3}{1\ mol\ Fe2O3} = 47.3\ g\ Fe_2O_3$$

FeS_2 is the limiting reactant.

8. How many grams of excess reactant are left over?

$$36.0\ g\ of\ FeS_2 \times \frac{1\ mol\ FeS}{120\ g\ Fes} \times \frac{11\ mol\ O2}{4\ mol\ FeS} \times \frac{32\ g\ O2}{1\ mol\ O2} = 26.4\ g\ O_2\ used$$

52.2 g O_2 available - 26.4 g O_2 used = 25.8 g O_2 left over after the reaction

9. How many grams of each product can be formed?

$$36.0\ g\ of\ FeS_2 \times \frac{1\ mol\ FeS}{120\ g\ Fes} \times \frac{2\ mol\ Fe2O3}{4\ mol\ FeS} \times \frac{159.6\ g\ Fe2O3}{1\ mol\ Fe2O3} = 23.9\ g\ Fe_2O_3$$

$$36.0\ g\ of\ FeS_2 \times \frac{1\ mol\ FeS}{120\ g\ Fes} \times \frac{8\ mol\ SO2}{4\ mol\ FeS} \times \frac{64.1\ g\ SO2}{1\ mol\ SO2} = 38.5\ g\ SO_2$$

10. If 15.6 g of Fe_2O_3 are formed, what is the percent yield for the reaction? (write the percent yield formula, then solve. Report % to 2 decimals)

$$\%\ yield = \frac{actual\ yield}{theoretical\ yield} \times 100 = \qquad \%\ yield = \frac{15.6\ g\ Fe2O3}{23.9\ g\ Fe2O3} \times 100 = 65.27\%$$

11. If the percent yield for the reaction was 35.88%, what mass of SO_2 could form? (write the correct version of the % yield formula, then solve. Report answer to sig figs.)

$$Actual\ Yield = \frac{\%\ yield \times theoretical\ yield}{100} = \qquad Actual\ Yield = \frac{35.88\% \times 38.5\ g\ SO2}{100} = 13.8\ g\ SO_2$$

Name: ______________________________ Date: ______________ Period: ______

Worksheet 14: Stoichiometry Mixed Self-Test 2

Estimated time: 40 minutes

Directions: Calculate the molar masses. Balance the equation and use it to answer the questions. Watch sig figs. Identify the type of stoichiometry involved. Include all units and labels.

$$\_\_\_\_ C_2H_6 + \_\_\_\_ O_2 \rightarrow \_\_\_\_ CO_2 + \_\_\_\_ H_2O$$

C_2H_6 =	O_2 =	CO_2 =	H_2O =

1. How many moles of CO_2 would form if 0.058 moles of C_2H_6 reacted?
Type:

2. How many grams of H_2O would be formed if 525 g of CO_2 were also formed?
Type:

3. How many grams of C_2H_6 are needed to fully react 0.582 moles of O_2?
Type:

4. How many moles of CO_2 would be formed from fully reacting 275 g of C_2H_6?
Type:

5. How many moles of water would form from a reaction of 9.3 moles of O_2?
Type:

6. If one started with 3555 g of C_2H_6, how many grams of H_2O would form?
Type:

Directions: Calculate the molar masses. Then, balance the equation and solve the problems.

$$\_\_\_\_ Hg(OH)_2 + \_\_\_\_ H_3PO_4 \rightarrow \_\_\_\_ Hg_3(PO_4)_2 + \_\_\_\_ H_2O$$

$Hg(OH)_2$ =	H_3PO_4 =	$Hg_3(PO_4)_2$ =	H_2O =

If given 5.00 g of each reactant ...

7. What is the limiting reactant?

8. How many grams of excess reactant are left over?

9. How many grams of each product can be formed?

10. If 4.32 g of $Hg(PO_4)_2$ are formed, what is the percent yield for the reaction? (write the percent yield formula, then solve. Report % to 2 decimals)

11. If the percent yield for the reaction was 93.75%, what mass of H_2O could form? (write the correct version of the % yield formula, then solve. Report answer to sig figs.)

Name: ______________________ Date: ______________ Period: ______

Worksheet 14 Key: Stoichiometry Mixed Self-Test 2

Estimated time: 40 minutes

Directions: Calculate the molar masses. Balance the equation and use it to answer the questions. Watch sig figs. Identify the type of stoichiometry involved. Include all units and labels.

$$2\ C_2H_6 + 7\ O_2 \rightarrow 4\ CO_2 + 6\ H_2O$$

$C_2H_6 = \frac{30\ g\ C2H6}{1\ mol\ C2H6}$	$O_2 = \frac{32\ g\ O2}{1\ mol\ O2}$	$CO_2 = \frac{44\ g\ CO2}{1\ mol\ CO2}$	$H_2O = \frac{18\ g\ H2O}{1\ mol\ H2O}$

1. How many moles of CO_2 would form if 0.058 moles of C_2H_6 reacted?

Type: moles to moles

$.058$ moles C_2H_6 x $\frac{4\ mol\ CO2}{2\ mol\ C2H6}$ = 0.12 mol CO_2

2. How many grams of H_2O would be formed if 525 g of CO_2 were also formed?

Type: mass to mass

525 g CO_2 x $\frac{1\ mol\ CO2}{44\ g\ CO2}$ x $\frac{6\ mol\ H2O}{4\ mol\ CO2}$ x $\frac{18\ g\ H2O}{1\ mol\ H2O}$ = 322 g H_2O

3. How many grams of C_2H_6 are needed to fully react 0.582 moles of O_2?

Type: moles to mass

$.582$ moles O_2 x $\frac{2\ mol\ C2H6}{7\ mol\ O2}$ x $\frac{30\ g\ C2H6}{1\ mol\ C2H6}$ = 4.99 g C_2H_6

4. How many moles of CO_2 would be formed from fully reacting 275 g of C_2H_6?

Type: mass to moles

275 g C_2H_6 x $\frac{1\ mol\ C2H6}{30\ g\ C2H6}$ x $\frac{4\ mol\ CO2}{2\ mol\ C2H6}$ = 18.3 mol CO_2

5. How many moles of water would form from a reaction of 9.3 moles of O_2?

Type: moles to moles

9.3 moles O_2 x $\frac{6\ mol\ H2O}{7\ mol\ O2}$ = 8.0 mol H_2O

6. If one started with 3555 g of C_2H_6, how many grams of H_2O would form?

Type: mass to mass

3555 g C_2H_6 x $\frac{1\ mol\ C2H6}{30\ g\ C2H6}$ x $\frac{6\ mol\ H2O}{2\ mol\ C2H6}$ x $\frac{18\ g\ H2O}{1\ mol\ H2O}$ = 6399 g H_2O

Name: ______________________________ Date: _____________ Period: ______

Worksheet 9: Limiting Reactant Stoichiometry Introduction

Directions: Balance the following equation and use it to answer the questions that follow. *Hint: 2 and 3 start with the limiting reactant.*

____ Al (s) + ____ HCl (aq) → ____ $AlCl_3$ (aq) + ____ H_2 (g)

Al = $\frac{27\ g\ Al}{1\ mol\ Al}$	HCl = $\frac{36.5\ g\ HCl}{1\ mol\ HCl}$	$AlCl_3$ = $\frac{133.5\ g\ AlCl3}{1\ mol\ AlCl3}$	H_2 = $\frac{2.0\ g\ H2}{1\ mol\ H2}$

If you have 29.8 g of Al reacting with 8.05 g of HCl ...

1. Which is the limiting reactant? *(**Helpful Hint using method 2**: Use molar mass to change to reactant moles; then use mole ratios obtained from the balanced equation to find out how many moles of one product you could find. Please note, that it has to be the SAME product in both sets of conversions. The one that could produce less of the product is the limiting reactant. **Caution**: the mass of product is NOT the answer, your answer is the reactant that leads to the least product. **Note**: there is another way to do this problem)*

2. How many grams of the excess reactant are left over after the reaction takes place? *(**Hint:** find the amount of grams of the excess are used and subtract that from the number of grams available)*

3. In theory, how many grams of each product would be formed? *(**Hint**: Get moles of the limiting reactant. Use mole ratio b/t LR and the desired product. Change from mol to grams product.)*

4. If 6.32 g $AlCl_3$ are formed, what is the percent yield for the reaction? (2 decimal places)
*(**Hint:** % yield = (actual yield/ theoretical yield) x 100)*

5. If the same percent yield as problem 4 holds true, how many g of H_2 will be produced?
*(**Hint:** actual yield = (theoretical yield x % yield) / 100)*

Name: ______________________________ Date: ______________ Period: ______

Worksheet 9 Key: Limiting Reactant Stoichiometry Introduction

Directions: Balance the following equation and use it to answer the questions that follow. *Hint: 2 and 3 start with the limiting reactant.*

$$2\ Al\ (s) + 6\ HCl\ (aq) \rightarrow 2\ AlCl_3\ (aq) + 3\ H_2\ (g)$$

$Al = \frac{27\ g\ Al}{1\ mol\ Al}$	$HCl = \frac{36.5\ g\ HCl}{1\ mol\ HCl}$	$AlCl_3 = \frac{133.5\ g\ AlCl3}{1\ mol\ AlCl3}$	$H_2 = \frac{2.0\ g\ H2}{1\ mol\ H2}$

If you have 29.8 g of Al reacting with 8.05 g of HCl ...

1. Which is the limiting reactant? *(**Helpful Hint using method 2**: Use molar mass to change to reactant moles; then use mole ratios obtained from the balanced equation to find out how many moles of one product you could find. Please note, that it has to be the SAME product in both sets of conversions. The one that could produce less of the product is the limiting reactant. **Caution**: the mass of product is NOT the answer, your answer is the reactant that leads to the least product. **Note**: there is another way to do this problem)*

Method 2:

$$29.8\ g\ Al \times \frac{1\ mol\ Al}{27\ g\ Al} \times \frac{2\ mol\ AlCl3}{2\ mol\ Al} = 1.10\ mol\ AlCl_3$$

$$8.05\ g\ HCl \times \frac{1\ mol\ HCl}{36.5\ g\ HCl} \times \frac{2\ mol\ AlCl3}{6\ mol\ HCl} = 0.0735\ mol\ AlCl_3$$

HCl is the limiting reactant.

Explanation: Method 2 assumes that both are the LR, knowing that only one of those statements can be true. If Al is the LR, then I should be able to get 1.10 mol $AlCl_3$. If HCl is the LR, then the max amount of $AlCl_3$ that can be formed is 0.0735 moles $AlCl_3$. If these are happening at the same time, then only the second statement can be true because the reaction would stop once the HCl runs out.

Method 1:

$$29.8\ g\ Al \times \frac{1\ mol\ Al}{27\ g\ Al} \times \frac{6\ mol\ HCl}{2\ mol\ Al} \times \frac{36.5\ g\ HCl}{1\ mol\ HCl} = 121\ g\ HCl$$

Explanation: Method one assumes one reactant is limiting and sees how much of the other is needed. If Al is the LR, I need 121 g HCl. Since I only have 8.05 g HCl, I guessed wrong. HCl must have run out first.

$$2\ Al\ (s) + 6\ HCl\ (aq) \rightarrow 2\ AlCl_3\ (aq) + 3\ H_2\ (g)$$

2. How many grams of the excess reactant are left over after the reaction takes place? ***(Hint:** find the amount of grams of the excess are used and subtract that from the number of grams available)*

$$8.05\text{ g HCl} \times \frac{1\ mol\ \text{HCl}}{36.5\ g\ \text{HCl}} \times \frac{2\ mol\ Al}{6\ mol\ HCl} \times \frac{27\ g\ Al}{1\ mol\ Al} = 1.98\text{ g Al used}$$

29.8 g of Al available
\- 1.98 g Al used
27.8 g Al are left over

Explanation: The first part of this involves converting from g of the LR to g of the nonLR. Stick with those sig figs. After that, you need to subtract the amount of Al you calculated were used in the reaction from the available amount of g of Al. The sig figs in these problems can be tricky. For the subtraction part, the sig figs are controlled by the number of decimal places. Since the given amount only had 1 decimal, the answer should only have 1 decimal.

3. In theory, how many grams of each product would be formed? ***(Hint**: Get moles of the limiting reactant. Use mole ratio b/t LR and the desired product. Change from mol to grams product.)*

$$8.05\text{ g HCl} \times \frac{1\ mol\ \text{HCl}}{36.5\ g\ \text{HCl}} \times \frac{2\ mol\ \text{AlCl3}}{6\ mol\ HCl} \times \frac{133.5\ g\ \text{AlCl3}}{1\ mol\ \text{AlCl3}} = 9.81\text{ g } AlCl_3$$

$$8.05\text{ g HCl} \times \frac{1\ mol\ \text{HCl}}{36.5\ g\ \text{HCl}} \times \frac{3\ mol\ \text{H2}}{6\ mol\ HCl} \times \frac{2\ g\ \text{H2}}{1\ mol\ \text{H2}} = 0.221\text{ g } H_2$$

4. If 6.32 g $AlCl_3$ are formed, what is the percent yield for the reaction? (2 decimal places) ***(Hint:** % yield = (actual yield/ theoretical yield) x 100)*

$$\%\text{ yield} = \frac{actual\ yield}{theoretical\ yield} \times 100\%$$

$$\%\text{ yield} = \frac{6.32\text{ g AlCl3}}{9.81\text{ g AlCl3}} \times 100\% = 64.42\%$$

5. If the same percent yield as problem 4 holds true, how many g of H_2 will be produced? ***(Hint:** actual yield = (theoretical yield x % yield) / 100)*

$$\text{Actual yield} = \frac{\%\text{ yield x theoretical yield}}{100}$$

$$\text{Actual yield} = \frac{64.42\%\ \text{x}\ 0.221\text{ g H2}}{100} = 0.142\text{ g } H_2$$

The other way ...

$$\%\text{ yield} = \frac{actual\ yield}{theoretical\ yield} \times 100\%$$

$$64.42\% = \frac{actual\ yield}{.221\ g\ H2} \times 100\%$$

Explanation: To solve, you need to multiply both sides by .221 g H_2. To undo the x 100 part, you'd need to divide both sides by 100.

Name: ________________________________ Date: ______________ Period: ______

Worksheet 10: Limiting Reactant Stoichiometry Practice

Directions: Use the conversion factor method. You MUST use units and labels on everything! Balance the equation and then use it to answer the questions that follow.

____ NH_3 (g) + ____ O_2 (g) → ____ NO (g) + ____ H_2O (g)

$NH_3 = \frac{17\ g\ NH3}{1\ mol\ NH3}$	$O_2 = \frac{32\ g\ O2}{1\ mol\ O2}$	$NO = \frac{30\ g\ NO}{1\ mol\ NO}$	$H_2O = \frac{18\ g\ H2O}{1\ mol\ H2O}$

If you have 136 g of NH_3 and 387 g of O_2 ...

1. Which is the limiting reactant?

2. How many grams of the excess reactant are left over?

3. How many grams of each product would be produced in an ideal reaction? (these are the theoretical yields)

4. If 202 g of NO were produced, what would the reaction's percent yield be? (show answer to 2 decimal places; be sure to show formula)

5. If the reaction had a percent yield of 76.58%, how many grams of water would be produced? (write the formula; watch sig figs)

Name: ______________________________ Date: _____________ Period: ______

Worksheet 10 Key: Limiting Reactant Stoichiometry Practice

Directions: Use the conversion factor method. You MUST use units and labels on everything! Balance the equation and then use it to answer the questions that follow.

$$4\ NH_3\ (g) + 5\ O_2\ (g) \rightarrow 4\ NO\ (g) + 6\ H_2O\ (g)$$

$NH_3 = \frac{17\ g\ NH3}{1\ mol\ NH3}$	$O_2 = \frac{32\ g\ O2}{1\ mol\ O2}$	$NO = \frac{30\ g\ NO}{1\ mol\ NO}$	$H_2O = \frac{18\ g\ H2O}{1\ mol\ H2O}$

If you have 136 g of NH_3 and 387 g of O_2 ...

1. Which is the limiting reactant?

$$136\ g\ NH_3 \times \frac{1\ mol\ NH3}{17\ g\ NH3} \times \frac{4\ mol\ NO}{4\ mol\ NH3} = 8.00\ mol\ NO$$

$$387\ g\ O_2 \times \frac{1\ mol\ O2}{32\ g\ O2} \times \frac{4\ mol\ NO}{5\ mol\ O2} = 9.68\ mol\ NO$$

NH_3 is the LR.

Explanation: I chose method 2 (going to the same product). NH_3 is the LR because 8.00 moles of NO is less than 9.68 moles of NO. The LR is the reactant that results in the smaller amount of product.

Let's say I tried method 1 (starting with one at random and changing to the other reactant).

$$136\ g\ NH_3 \times \frac{1\ mol\ NH3}{17\ g\ NH3} \times \frac{5\ mol\ O2}{4\ mol\ NH3} \times \frac{32\ g\ O2}{1\ mol\ O2} = 320.\ g\ O_2$$

Explanation: If NH_3 is the LR, then I'd need 320. g of O_2 to fully react it. The question I need to ask myself is: do I have enough O_2? Because I was given 387 g O_2 to work with, the answer is "yes" I have enough. Therefore, my guess was correct, and NH_3 is indeed the limiting reactant.

Worksheet 10 Key continued: Limiting Reactant Stoichiometry Practice

$$4\ NH_3\ (g) + 5\ O_2\ (g) \rightarrow 4\ NO\ (g) + 6\ H_2O\ (g)$$

2. How many grams of the excess reactant are left over?

$$136\ g\ NH_3 \times \frac{1\ mol\ NH3}{17\ g\ NH3} \times \frac{5\ mol\ O2}{4\ mol\ NH3} \times \frac{32\ g\ O2}{1\ mol\ O2} = 320.\ g\ O_2\ \text{used}$$

387 g O_2 available – 320. g O_2 used = 67 g O_2 excess

Explanation: Start with the given amount of the LR, 136 g NH_3. Use the molar mass with the grams on the bottom and the moles of top to convert to moles of NH_3 $(\frac{1\ mol\ NH3}{17\ g\ NH3})$. Use the mole ratio between the LR and the nonLR to get to moles of O_2. The mole ratio looks like this $(\frac{5\ mol\ O2}{4\ mol\ NH3})$. Use the molar mass of the nonLR $(\frac{32\ g\ O2}{1\ mol\ O2})$ to get to the mass of the O_2 used in the reaction. The answer to 3 significant figures is 320. g O_2. Keep in mind what this is saying. If I used all of the LR, I would use 320. g of the nonLR. The question asks about the grams of the nonLR that are in excess (left over). So, to get this, I take the given amount of the nonLR, which is 387 g of O_2, and subtract the amount used in the reaction, 320. g O_2. The answer adjusted for significant figures is 67 g O_2.

3. How many grams of each product would be produced in an ideal reaction? (these are the theoretical yields)

$$136\ g\ NH_3 \times \frac{1\ mol\ NH3}{17\ g\ NH3} \times \frac{4\ mol\ NO}{4\ mol\ NH3} \times \frac{30\ g\ NO}{1\ mol\ NO} = 240.\ g\ NO$$

$$136\ g\ NH_3 \times \frac{1\ mol\ NH3}{17\ g\ NH3} \times \frac{6\ mol\ H2O}{4\ mol\ NH3} \times \frac{18\ g\ H2O}{1\ mol\ H2O} = 216\ g\ H_2O$$

Explanation:

- Start with the given amount of the LR: 136 g NH_3.
- In both sets of conversions, you divide by the 17 g of the NH_3 and multiply by 1 mol NH_3 to get to moles of the LR.
- Switch to moles of the products. Since there are two products, there are two distinct sets of conversions. One with a mole ratio of $(\frac{4\ mol\ NO}{4\ mol\ NH3})$, and the other with a mole ratio of $(\frac{6\ mol\ H2O}{4\ mol\ NH3})$.
- Use the molar mass of each product to convert from grams to moles. Again, we have two products, so there are two sets of conversions. The first product is NO, so its conversion looks like this: $\frac{30\ g\ NO}{1\ mol\ NO}$. The second product is H_2O, so it ends with $\frac{18\ g\ H2O}{1\ mol\ H2O}$.
- Solve both conversion sets by multiplying the LR by the top values and dividing all bottom values.
- Report answers to three sig figs. The maximum amount of NO that can be formed is 240. g NO. The max amount of H_2O that can be formed is 216 g H_2O.

Worksheet 10 Key continued: Limiting Reactant Stoichiometry Practice

4. If 202 g of NO were produced, what would the reaction's percent yield be? (show answer to 2 decimal places; be sure to show formula)

$$\% \text{ yield} = \frac{actual\ yield}{theoretical\ yield} \times 100\% \qquad \% \text{ yield} = \frac{202\ g\ NO}{240.g\ NO} \times 100\% = 84.17\%$$

Explanation: Percent calculations involve dividing a part of something by the whole and multiplying by 100. In this case, the part that's formed is the actual yield in the problem. The theoretical yield is the maximum amount of that product that could be formed in a perfect reaction. A perfect reaction would have 100% yield. If 202 g NO were formed but 240. g NO should have been theoretically formed, the first step is to divide these two values $\frac{202\ g\ NO}{240\ g\ NO}$. Multiply by 100 to turn this decimal into a percent. I tell my students to go to 2 decimal places for all percents, but your teacher may require something different.

5. If the reaction had a percent yield of 76.58%, how many grams of water would be produced? (write the formula; watch sig figs)

$$\text{Actual yield} = \frac{\%\ \text{yield x theoretical yield}}{100} \qquad \text{Actual yield} = \frac{76.58\%\ \text{ x } 216\ \text{g H2O}}{100} = 165\ \text{g}\ H_2O$$

Explanation:

This is the same formula as the previous problem. It only looks different because it's rearranged to solve for the actual yield.

- To find actual yield, take the percent yield given, which is 76.58% and multiply by the theoretical amount of the desired product, in this case 216 g H_2O.
- To undo the percent portion, you need to divide by 100.
- The answer to 3 sig figs is 165 g H_2O.

Name: ________________________________ Date: ______________ Period: ______

Worksheet 11: Limiting Reactant Stoichiometry Self-Quiz 1

Directions: Use the conversion factor method. Use units and labels on everything! Balance the equation and then use it to answer the questions that follow.

$$\_\_\_\_ NH_3\ (g) + \_\_\_\_ O_2\ (g) \rightarrow \_\_\_\_ NO\ (g) + \_\_\_\_ H_2O\ (g)$$

$NH_3 = \frac{17\ g\ NH3}{1\ mol\ NH3}$	$O_2 = \frac{32\ g\ O2}{1\ mol\ O2}$	$NO = \frac{30\ g\ NO}{1\ mol\ NO}$	$H_2O = \frac{18\ g\ H2O}{1\ mol\ H2O}$

If you have 101 g of NH_3 and 5.87 g of O_2 ...

1. Which is the limiting reactant?

2. How many grams of the excess reactant are left over?

3. How many grams of each product would be produced in an ideal reaction?

4. If 1.52 g of NO were produced, what would the reaction's percent yield be? (show answer to 2 decimal places; be sure to show formula)

5. If the same percent yield held true, how many grams of water would also be produced? (write down the correct version of the formula, go to sig figs)

Name: ______________________________ Date: ______________ Period: ______

Worksheet 11 Key: Limiting Reactant Stoichiometry Self-Quiz 1

Directions: Use the conversion factor method. Use units and labels on everything! Balance the equation and then use it to answer the questions that follow.

$$4\ NH_3\ (g) + 5\ O_2\ (g) \rightarrow 4\ NO\ (g) + 6\ H_2O\ (g)$$

$NH_3 = \frac{17\ g\ NH3}{1\ mol\ NH3}$	$O_2 = \frac{32\ g\ O2}{1\ mol\ O2}$	$NO = \frac{30\ g\ NO}{1\ mol\ NO}$	$H_2O = \frac{18\ g\ H2O}{1\ mol\ H2O}$

If you have 101 g of NH_3 and 5.87 g of O_2 ...

1. Which is the limiting reactant?

Method 1:	**Method 2:**
$101\ g\ NH_3 \times \frac{1\ mol\ NH3}{17\ g\ NH3} \times \frac{5\ mol\ O2}{4\ mol\ NH3} \times \frac{32\ g\ O2}{1\ mol\ O2} = 238\ g\ O_2$	$101\ g\ NH_3 \times \frac{1\ mol\ NH3}{17\ g\ NH3} \times \frac{4\ mol\ NO}{4\ mol\ NH3} = 5.94\ mol\ NO$
OR	
$5.87\ g\ O_2 \times \frac{1\ mol\ O2}{32\ g\ O2} \times \frac{4\ mol\ NH3}{5\ mol\ O2} \times \frac{17\ g\ NH3}{1\ mol\ NH3} = 2.49\ g\ NH_3$	$5.87\ g\ of\ O_2 \times \frac{1\ mol\ O2}{32\ g\ O2} \times \frac{4\ mol\ NO}{5\ mol\ O2} = 0.147\ mol\ NO$

Conclusion: O_2 is the LR.

2. How many grams of the excess reactant are left over?

$$5.87\ g\ of\ O_2 \times \frac{1\ mol\ O2}{32\ g\ O2} \times \frac{4\ mol\ NH3}{5\ mol\ O2} \times \frac{17\ g\ NH3}{1\ mol\ NH3} = 2.49\ g\ NH_3\ used$$

101 g NH_3 available – 2.49 g NH_3 used = 99 g NH_3

3. How many grams of each product would be produced in an ideal reaction?

$$5.87\ g\ of\ O_2 \times \frac{1\ mol\ O2}{32\ g\ O2} \times \frac{4\ mol\ NO}{5\ mol\ O2} \times \frac{30\ g\ NO}{1\ mol\ NO} = 4.40\ g\ NO$$

$$5.87\ g\ of\ O_2 \times \frac{1\ mol\ O2}{32\ g\ O2} \times \frac{6\ mol\ H2O}{5\ mol\ O2} \times \frac{18\ g\ H2O}{1\ mol\ H2O} = 3.96\ g\ H_2O$$

4. If 1.52 g of NO were produced, what would the reaction's percent yield be? (show answer to 2 decimal places; be sure to show formula)

$$\%\ yield = \frac{actual\ yield}{theoretical\ yield} \times 100\% \qquad \%\ yield = \frac{1.52\ g\ NO}{4.40\ g\ NO} \times 100\% = 35.55\%$$

5. If the same percent yield held true, how many grams of water would also be produced? (write down the correct version of the formula, go to sig figs)

$$Actual\ yield = \frac{\%\ yield \times theoretical\ yield}{100} \qquad Actual\ yield = \frac{35.55\% \times 3.96\ g\ H2O}{100} = 1.41\ g\ H_2O$$

Name: ______________________________ Date: ______________ Period: ______

Worksheet 12: Limiting Reactant Stoichiometry Self-Quiz 2

____ $HClO_4$ + ____ P_4O_{10} → ____ H_3PO_4 + ____ Cl_2O_7

Directions: Balance the equation and answer each question using the conversion factor method. Watch sig figs.

$HClO_4 = \frac{100.5\ g\ HClO4}{1\ mol\ HClO4}$	$P_4O_{10} = \frac{284\ g\ P4O10}{1\ mol\ P4O10}$	$H_3PO_4 = \frac{98\ g\ H3PO4}{1\ mol\ H3PO4}$	$Cl_2O_7 = \frac{183\ g\ Cl2O7}{1\ mol\ Cl2O7}$

If given 2.54 g $HClO_4$ and 12.2 g P_4O_{10} ...

1. What is the limiting reactant?

2. What mass (in grams) of the excess reactant left over?

3. How many grams of each product can be formed?

4. If 0.698 g of H_3PO_4 are actually formed, what is the percent yield? (ans to 2 decimals)

5. Assuming the same percent yield hold true, how many grams of Cl_2O_7 will actually be formed?

Name: ______________________________ Date: ______________ Period: ______

Worksheet 12 Key: Limiting Reactant Stoichiometry Self-Quiz 2

$$12\ HClO_4 + P_4O_{10} \rightarrow 4\ H_3PO_4 + 6\ Cl_2O_7$$

Directions: Balance the equation and answer each question using the conversion factor method. Watch sig figs.

$HClO_4 = \frac{100.5\ g\ HClO4}{1\ mol\ HClO4}$	$P_4O_{10} = \frac{284\ g\ P4O10}{1\ mol\ P4O10}$	$H_3PO_4 = \frac{98\ g\ H3PO4}{1\ mol\ H3PO4}$	$Cl_2O_7 = \frac{183\ g\ Cl2O7}{1\ mol\ Cl2O7}$

If given 2.54 g $HClO_4$ and 12.2 g P_4O_{10} ...

1. What is the limiting reactant?

$$2.54\ g\ HClO_4 \times \frac{1\ mol\ HClO4}{100.5\ g\ HClO4} \times \frac{4\ mol\ H3PO4}{12\ mol\ HClO4} \times \frac{98\ g\ H3PO4}{1\ mol\ H3PO4} = 0.826\ g\ H_3PO_4$$

$$12.2\ g\ P_4O_{10} \times \frac{1\ mol\ P4O10}{284\ g\ P4O10} \times \frac{4\ mol\ H3PO4}{1\ mol\ P4O10} \times \frac{98\ g\ H3PO4}{1\ mol\ H3PO4} = 16.8\ g\ H_3PO_4$$

$HClO_4$ is the Limiting Reactant

2. What mass (in grams) of the excess reactant left over?

$$2.54\ g\ HClO_4 \times \frac{1\ mol\ HClO4}{100.5\ g\ HClO4} \times \frac{1\ mol\ P4O10}{12\ mol\ HClO4} \times \frac{284\ g\ P4O10}{1\ mol\ P4O10} = 0.598\ g\ P_4O_{10}$$

12.2 g P_4O_{10} available - 0.598 g P_4O_{10} used = 11.6 g P_4O_{10} left over

3. How many grams of each product can be formed?

$$2.54\ g\ HClO_4 \times \frac{1\ mol\ HClO4}{100.5\ g\ HClO4} \times \frac{4\ mol\ H3PO4}{12\ mol\ HClO4} \times \frac{98\ g\ H3PO4}{1\ mol\ H3PO4} = 0.826\ g\ H_3PO_4$$

$$2.54\ g\ HClO_4 \times \frac{1\ mol\ HClO4}{100.5\ g\ HClO4} \times \frac{6\ mol\ Cl2O7}{12\ mol\ HClO4} \times \frac{183\ g\ Cl2O7}{1\ mol\ Cl2O7} = 2.31\ g\ Cl_2O_7$$

4. If 0.698 g of H_3PO_4 are actually formed, what is the percent yield? (ans to 2 decimals)

$$\%\ yield = \frac{actual\ yield}{theoretical\ yield} \times 100 \qquad \%\ yield = \frac{.698\ g\ H3PO4}{.826\ g\ H3PO4} \times 100 = 84.50\%$$

5. Assuming the same percent yield hold true, how many grams of Cl_2O_7 will actually be formed?

$$Actual\ yield = \frac{(\%\ yield)\ (theoretical\ yield)}{100} \qquad Actual\ yield = \frac{(84.50\%)\ (2.31\ g\ Cl2O7)}{100} = 1.95\ g\ Cl_2O_7$$

Timed Tests Rationale:

The next few worksheets are timed self-tests. Many chemistry topics take a lot of practice. It's great to see videos and hear people explain how to do the problems. That said, there's really no substitute for working through the problems on your own.

Hints/tips/tricks:

- Every number should be followed by a unit and a chemical label. I know it gets tedious, but it will help you track the chemicals as you go through the problems.
- Use the calculator, pen/pencil, and periodic table you would normally use for a formal assessment. The more comfortable you are with the tools at your disposal, the easier it will be.
- Set a stopwatch and then turn it upside down so you can't watch it. Work through the problems at your natural pace to get a baseline for how quickly you work. Weigh this mentally against how many questions you know your teacher will present you. If there's time, redo a sheet a day later to see if you can improve your time.

Name: ________________________________ Date: _______________ Period: ______

Worksheet 13: Stoichiometry Mixed Self-Test 1

Estimated time: 40 minutes

Directions: Balance the equation and use it to answer the questions. Watch sig figs. Identify the type of stoichiometry involved. Include all units and labels.

____ As + ____ NaOH → ____ Na_3AsO_3 + ____ H_2

As = $\frac{74.9\ g\ As}{1\ mol\ As}$	NaOH = $\frac{40\ g\ NaOH}{1\ mol\ NaOH}$	Na_3AsO_3 = $\frac{191.9\ g\ Na3AsO3}{1\ mol\ Na3AsO3}$	H_2= $\frac{2\ g\ H2}{1\ mol\ H2}$

1. If 352 g of H_2 were formed, how many moles of NaOH must have reacted?
Type:

2. If 64 g of As reacted, how many grams of NaOH also reacted?
Type:

3. If 0.58 moles of Na_3AsO_3 were formed, how many grams of H_2 would also be formed?
Type:

4. If 6.8 x 10^{43} moles of NaOH reacted, what mass of As must also be reacting?
Type:

5. How many moles of NaOH must react to form 54.9 moles of Na_3AsO_3?
Type:

6. How many grams of Na_3AsO_3 would form in a reaction involving 7.99 moles of NaOH?
Type:

Directions: Calculate the molar masses. Then, balance the equation and solve the problems.

$$\_\_\_\_ FeS_2 + \_\_\_\_ O_2 \rightarrow \_\_\_\_ Fe_2O_3 + \_\_\_\_ SO_2$$

FeS_2 =	O_2 =	Fe_2O_3 =	SO_2 =

If given 36.0 g of FeS_2 and 52.2 g of O_2 ...

7. What is the limiting reactant?

8. How many grams of excess reactant are left over?

9. How many grams of each product can be formed?

10. If 15.6 g of Fe_2O_3 are formed, what is the percent yield for the reaction? (write the percent yield formula, then solve. Report % to 2 decimals)

11. If the percent yield for the reaction was 35.88%, what mass of SO_2 could form? (write the correct version of the % yield formula, then solve. Report answer to sig figs.)

Name: ______________________________ Date: ______________ Period: _____

Worksheet 13 Key: Stoichiometry Mixed Self-Test 1

Estimated time: 40 minutes

Directions: Balance the equation and use it to answer the questions. Watch sig figs. Identify the type of stoichiometry involved. Include all units and labels.

$$2\ As + 6\ NaOH \rightarrow 2\ Na_3AsO_3 + 3\ H_2$$

$As = \frac{74.9\ g\ As}{1\ mol\ As}$	$NaOH = \frac{40\ g\ NaOH}{1\ mol\ NaOH}$	$Na_3AsO_3 = \frac{191.9\ g\ Na3AsO3}{1\ mol\ Na3AsO3}$	$H_2 = \frac{2\ g\ H2}{1\ mol\ H2}$

1. If 352 g of H_2 were formed, how many moles of NaOH must have reacted?
Type: mass to moles

$352\ g\ of\ H_2 \times \frac{1\ mol\ H2}{2\ g\ H2} \times \frac{6\ mol\ NaOH}{3\ mol\ H2} = 352\ mol\ NaOH$

2. If 64 g of As reacted, how many grams of NaOH also reacted?
Type: mass to mass

$64\ g\ As \times \frac{1\ mol\ As}{74.9\ g\ As} \times \frac{6\ mol\ NaOH}{2\ mol\ As} \times \frac{40\ g\ NaOH}{1\ mol\ NaOH} = 1.0 \times 10^2\ g\ NaOH$ or 100 g NaOH with a line over the first 0

3. If 0.58 moles of Na_3AsO_3 were formed, how many grams of H_2 would also be formed?
Type: moles to mass

$.58\ moles\ Na_3AsO_3 \times \frac{3\ mol\ H2}{2\ mol\ Na3AsO3} \times \frac{2\ g\ H2}{1\ mol\ H2} = 1.7\ g\ H_2$

4. If 6.8×10^{43} moles of NaOH reacted, what mass of As must also be reacting?
Type: moles to mass

$6.8 \times 10^{43}\ moles\ NaOH \times \frac{2\ mol\ As}{6\ mol\ NaOH} \times \frac{74.9\ g\ As}{1\ mol\ As} = 1.7 \times 10^{45}\ g\ As$

5. How many moles of NaOH must react to form 54.9 moles of Na_3AsO_3?
Type: moles to moles

$54.9\ moles\ Na_3AsO_3 \times \frac{6\ mol\ NaOH}{2\ mol\ Na3AsO3} = 165\ mol\ NaOH$

6. How many grams of Na_3AsO_3 would form in a reaction involving 7.99 moles of NaOH?
Type: moles to mass

$7.99\ moles\ NaOH \times \frac{2\ mol\ Na3AsO3}{6\ moles\ NaOH} \times \frac{191.9\ g\ Na3AsO3}{1\ mol\ Na3AsO3} = 511\ g\ Na_3AsO_3$

Directions: Calculate the molar masses. Then, balance the equation and solve the problems.

$$3\ Hg(OH)_2 + 2\ H_3PO_4 \rightarrow Hg_3(PO_4)_2 + 6\ H_2O$$

$Hg(OH)_2 = \frac{234.6\ g\ Hg(OH)2}{1\ mol\ Hg(OH)2}$	$H_3PO_4 = \frac{98\ g\ H3PO4}{1\ mol\ H3PO4}$	$Hg_3(PO_4)_2 = \frac{791.8\ g\ Hg3(PO4)2}{1\ mol\ Hg3(PO4)2}$	$H_2O = \frac{18\ g\ H2O}{1\ mol\ H2O}$

If given 5.00 g of each reactant …

7. What is the limiting reactant?

$$5.00\ g\ Hg(OH)_2 \times \frac{1\ mol\ Hg(OH)2}{234.6\ g\ Hg(OH)2} \times \frac{6\ mol\ H2O}{3\ mol\ Hg(OH)2} \times \frac{18\ g\ H2O}{1\ mol\ H2O} = 0.767\ g\ H_2O$$

$$5.00\ g\ H_3PO_4 \times \frac{1\ mol\ H3PO4}{98\ g\ H3PO4} \times \frac{6\ mol\ H2O}{2\ mol\ H3PO4} \times \frac{18\ g\ H2O}{1\ mol\ H2O} = 2.76\ g\ H_2O$$

$Hg(OH)_2$ is the limiting reactant.

8. How many grams of excess reactant are left over?

$$5.00\ g\ Hg(OH)_2 \times \frac{1\ mol\ Hg(OH)2}{234.6\ g\ Hg(OH)2} \times \frac{2\ mol\ H3PO4}{3\ mol\ Hg(OH)2} \times \frac{98\ g\ H3PO4}{1\ mol\ H3PO4} = 1.39\ g\ H_3PO_4 \text{ are used}$$

5.00 g H_3PO_4 available - 1.39 g H_3PO_4 used = 3.61 g H_3PO_4 left over after the reaction

9. How many grams of each product can be formed?

$$5.00\ g\ Hg(OH)_2 \times \frac{1\ mol\ Hg(OH)2}{234.6\ g\ Hg(OH)2} \times \frac{6\ mol\ H2O}{3\ mol\ Hg(OH)2} \times \frac{18\ g\ H2O}{1\ mol\ H2O} = 0.767\ g\ H_2O$$

$$5.00\ g\ Hg(OH)_2 \times \frac{1\ mol\ Hg(OH)2}{234.6\ g\ Hg(OH)2} \times \frac{1\ mol\ Hg3(PO4)2}{3\ mol\ Hg(OH)2} \times \frac{791.8\ g\ Hg3(PO4)2}{1\ mol\ Hg3(PO4)2} = 5.63\ g\ Hg_3(PO_4)_2$$

10. If 4.32 g of $Hg_3(PO_4)_2$ are formed, what is the percent yield for the reaction? (write the percent yield formula, then solve. Report % to 2 decimals)

$$\%\ yield = \frac{actual\ yield}{theoretical\ yield} \times 100 = \qquad \%\ yield = \frac{4.32\ g\ Hg3(PO4)2}{5.63\ g\ Hg3(PO4)2} \times 100 = 76.73\%$$

11. If the percent yield for the reaction was 93.75%, what mass of H_2O could form? (write the correct version of the % yield formula, then solve. Report answer to sig figs.)

$$Actual\ Yield = \frac{\%\ yield\ x\ theoretical\ yield}{100} = \qquad Actual\ Yield = \frac{93.75\%\ x\ 0.767\ g\ H2O}{100} = .719\ g\ H_2O$$

Thank You for Reading:

Chemistry takes a lot of practice. This workbook focused on Stoichiometry. Check out *5 Steps to Surviving Chemistry* if you're looking for a more comprehensive review of other chem topics. You may also wish to grab *5 Steps Chemistry Workbook Series Book 2: Mighty Mole Concepts*.

I hope this book helps you along your journey. If you have questions, please feel free to email me at devyaschildren@gmail.com.

If you're ready for a break from chem, visit my website (http://www.juliecgilbert.com/) to read up on some of my fiction series. Whether you're in the mood for a heavy hitter like Beyond Broken Pencils (about a school shooting) or a light YA scifi like Devya's Children, there's probably a story that will appeal to you.

If you're interested in virtual tutoring via zoom, send inquiries to the above email address.

Good luck.

Sincerely,

Julie C. Gilbert

Made in United States
Orlando, FL
04 November 2026

90208505R00036